Wildfowl 67

Editor
Eileen C. Rees

Associate Editor
Anthony D. Fox

GW00480723

Published by
Wildfowl & Wetlands Trust
Slimbridge, Gloucestershire
GL2 7BT, UK
Registered Charity No. 1030884

Wildfowl

Editor

Dr Eileen C. Rees
Wildfowl & Wetlands Trust
Slimbridge, Gloucestershire GL2 7BT, UK.

Associate Editor

Prof Anthony D. Fox
Department of Bioscience, Aarhus University
Kalø, Grenåvej 14, DK-8410 Rønde, Denmark.

Cover photograph: Snow Goose *Anser caerulescens* coming in to land at Bosque del Apache, New Mexico, USA. © Danny Green/naturepl.com

Cover design by Paul Marshall

Published by the Wildfowl & Wetlands Trust, Slimbridge, Gloucestershire GL2 7BT, UK

Wildfowl is available by subscription from the above address. For further information call +44 (0)1453 891900 (extension 257), or e-mail wildfowl@wwt.org.uk

ISBN 978 0 900806 71 1
Print ISSN 0954-6324
On-line ISSN 2052-6458

Printed on FSC™ certified paper by Henry Ling Limited, at the Dorset Press, DT1 1HD.

Contents

Short Communications

Photograph: A pair of Greenland White-fronted Geese flying in western Iceland during spring 2013, by Tony Fox.

Wildfowl 67: **Editorial**

Much of our current understanding of waterbird populations is based on long-term monitoring and research programmes, sometimes spanning several decades. It is therefore gratifying to see papers in the current issue of *Wildfowl* based on observations made by dedicated researchers over many years, including a review of recent analyses undertaken to determine reasons for the decline in Greenland White-fronted Goose *Anser albifrons flavirostris* numbers from 1999 onwards. Readers may be aware that the Greenlandic subspecies of the Greater White-fronted Goose was first described by Peter Scott with Christopher Dalgety in 1949, so it seems particularly appropriate to have a synthesis of current knowledge on the drivers of population trends for these birds presented in WWT's scientific journal. This particular paper focuses on geese wintering at Wexford in southeast Ireland, where a catching, marking and resighting programme has been underway since 1983, and indicates that a reduction in productivity in recent years may be an important reason underlying the population change. The authors noted however that further studies are required to confirm the reason for non-breeding or breeding failure once the birds reach Greenland, and also that the demography of other Greenland White-fronted Goose wintering flocks (especially those wintering on Islay, northwest Scotland, the second-largest wintering area for the subspecies) should be quantified to throw light on the demography of the population as a whole.

Long-term studies also provide an invaluable base upon which to develop informative shorter-term research programmes. As such, hypotheses regarding the morphology of Lesser Snow Geese *Anser caerulescens caerulescens* in southwest Louisiana, analysed and presented here, were derived from earlier studies of their different feeding habitats in this part of the wintering range. Surveys repeated at intervals also provide valuable insight into population change, although reasons for the change may be more difficult to identify. For instance, a comparison of the sex ratio of the Common Pochard *Aythya ferina* recorded across Europe and parts of North Africa in winter 2015/16 with results of a previous survey carried out over the same area in winters 1988/89–1989/90 (published in *Wildfowl* 46, accessible on-line at wildfowl.wwt.org.uk) found that, whilst there was a preponderance of males in both surveys, the proportion is significantly higher now than *c.* 25 years ago. In the same vein, monitoring at intervals of heterospecific brood parasitism (HBP) among duck species breeding in South Bohemia, Czech Republic, since the 1970s indicate that rates of HBP are lower in the 21st century than back in 1975–1980, when duck breeding populations in the region were higher. Interestingly, we also now learn that Tufted Duck seems to be the most suitable host species as well as the most successful of five Anatidae species considered in parasitising other birds' nests.

Wildfowl has a strong tradition of publishing research on endangered species, including for example on the Laysan Teal *Anas laysanensis* which is classed as Critically Endangered in the IUCN Red List of Threatened Species (www.iucnredlist.org). For instance, factors affecting the onset of breeding and nesting phenology for these birds on Laysan Island in the Hawaiian archipelago were described by Michelle Reynolds and co-authors 10 years ago in *Wildfowl* 57. Forty-two Laysan Teal were translocated from Laysan Island to Midway Atoll (also one of the Hawaiian islands) during 2004–2005, after which numbers increased to 661 individuals before dropping by 37% following an outbreak of botulism in 2015. A paper presented in the current issue of the journal considers methods for monitoring the abundance of Laysan Teal, and potentially other reintroduced species of conservation concern, which may be similarly susceptible to catastrophic population declines. Elsewhere, the island of Madagascar has three endemic duck species classified by IUCN as Endangered or Critically Endangered, and since 2010 several areas have been protected to secure key habitat for these birds, including the one site at which the Madagascar Pochard *Aythya innotata* occurs. Felix Razafindrajao and co-authors therefore present here the results of their modelling of bird and habitat distribution data for Madagascar Teal *Anas bernieri* and Meller's Duck *A. melleri*, to estimate the extent of suitable habitat for these endangered species, determine whether this coincides with areas already protected, and thus provide a focus for future survey work and site protection.

As always, I am hugely indebted to the great team of individuals involved in the production of *Wildfowl*. A higher number of papers than usual were submitted to this year's issue of the journal, so I am particularly grateful not only to the authors for considering *Wildfowl* for publication of their work but to the referees for giving their time to peer-review the manuscripts, sometimes at very short notice. Tony Fox (Associate Editor for *Wildfowl*) and Editorial Board members Jeff Black, Bruce Dugger, Andy Green and Matt Guillemain have provided vital support and sound scientific advice throughout the process. I also thank Ellen Matthews (EM Typesetting) for preparing the proofs, Paul Marshall for designing the cover, the staff at Henry Ling Ltd for the printed copies and Maggie Sage, Linda Dickerson and Jane Gawthorne-Dover for kindly providing administrative support. I hope that readers will be as interested as I am in the papers published in *Wildfowl* 67, and that the findings not only enhance your appreciation of the birds but also of the exceedingly diverse conditions that they encounter in different parts of the world.

Eileen Rees

Editor: *Wildfowl*
WWT Martin Mere

Diagnosing the decline of the Greenland White-fronted Goose *Anser albifrons flavirostris* using population and individual level techniques

MITCH D. WEEGMAN[1,2,6]*, ANTHONY DAVID FOX[3],
GEOFF M. HILTON[2], DAVID J. HODGSON[1], ALYN J. WALSH[4],
LARRY R. GRIFFIN[5] & STUART BEARHOP[1]

[1]Centre for Ecology and Conservation, College of Life and Environmental Sciences, University of Exeter, Cornwall Campus, Penryn TR10 9EZ, UK.
[2]Wildfowl & Wetlands Trust, Slimbridge, Gloucester GL2 7BT, UK.
[3]Department of Bioscience, Aarhus University, Kalø, Grenåvej 14, DK-8410 Rønde, Denmark.
[4]National Parks and Wildlife Service, Wexford Wildfowl Reserve, North Slob, Wexford, Republic of Ireland.
[5]Wildfowl & Wetlands Trust, Caerlaverock Wetland Centre, Eastpark Farm, Caerlaverock, Dumfriesshire DG1 4RS, UK.
[6]Present address: School of Natural Resources, University of Missouri, Columbia, Missouri, 65211 USA.
*Correspondence author. E-mail: weegmanm@missouri.edu

Abstract

Following an increase in numbers from 1982 to 1998, the Greenland White-fronted Goose *Anser albifrons flavirostris* declined over the period 1999–2015, stimulating detailed analyses at the population and individual level to provide a better understanding of the dynamics of this subspecies. Here we synthesise the results of the analyses in order to describe the potential reasons for the decline. Utilising a 27-year capture-mark-recapture dataset from the main wintering site for these birds (Wexford, Ireland), multistate models estimated sex-specific survival and movement probabilities. Our results suggested no evidence of a sex bias in emigration or "remigration" rates. These analyses formed the foundation for an integrated population model (IPM), which included population size and productivity data to assess source-sink dynamics of Wexford birds through estimation of age-, site-, and year-specific survival and movement probabilities. Results from the IPM suggested that Wexford is a large sink, and that a reduction in productivity is an important demographic mechanism underlying population change for birds wintering at the site. Low productivity may be due to environmental conditions in the breeding range, because birds bred successfully at youngest ages when conditions in Greenland were favourable in the year(s) during adulthood prior to and including the year of successful breeding. This effect could be mediated by prolonged parent-offspring relationships, as birds remained with parents into adulthood, forfeiting immediate reproductive success despite there being no fitness benefits to offspring of family associations after age 3 years. Global Positioning System and acceleration data collected from 15 male individuals suggested that two successful breeding birds were the only tagged individuals whose mate exhibited prolonged incubation. More data is required, however, to determine whether poor productivity is attributable to deferral of nesting or to failure of nesting attempts. Spring foraging did not appear to limit breeding or migration distance because breeding and non-breeding or failed-breeding birds, as well as Irish and Scottish birds, did not differ in their proportion of time spent feeding or on energy expenditure in spring. We recommend that future research should quantify the demography of other Greenland White-fronted Goose wintering flocks, to assess holistically the mechanisms underlying the global population decline.

Key words: animal movement, Global Positioning System-acceleration tracking devices, integrated population model, migratory birds, population decline.

Arctic-nesting geese are key species of northern hemisphere polar regions, acting as arctic ecosystem bioengineers through their grazing and grubbing of vegetation, and as important prey for other species (Bantle & Alisauskas 1998; Gauthier *et al.* 2004). Their conservation and management therefore is important to maintaining the integrity of arctic ecosystem functions. In recent decades, many goose populations around the world have increased, largely as a result of greater food availability associated with agricultural practices (Fox & Abraham 2017) and more informed management of hunting as a conservation tool (Owen 1990; Abraham & Jeffries 1997; Madsen *et al.* 1999; Gauthier *et al.* 2005). Typically, populations that remain of concern are those with limited ability to adapt to changing habitats or where hunting is uncontrolled. For example, the Red-breasted Goose *Branta ruficollis*, which is classed as Vulnerable by the International Union for Conservation of Nature (IUCN

2016), is believed to be hunted illegally particularly during migration in Kazakhstan and Russia, which has resulted in additive mortality and reduced population size in recent years (Cranswick *et al.* 2012). Likewise, a combination of decreased habitat availability and increased hunting in China of the Lesser White-fronted Goose *Anser erythropus*, also classed as Vulnerable globally by IUCN, is believed to have contributed to population decline (Wang *et al.* 2012). Nonetheless, there remain reasons to be cautious about the conservation and management of all arctic-nesting geese in future years, particularly with habitat changes associated with the warming climate, the resulting temperature increases of which are greatest at polar latitudes (IPCC 2014). These increases in temperature have already changed arctic ecosystems, contributing to greater variation in predator-prey interactions (Nolet *et al.* 2013) and "phenological mismatch" in food abundance as a result of differential changes in the onset of summer between temperate and polar regions (Durant *et al.* 2007; Tulp & Schekkerman 2008; Gilg *et al.* 2012).

To understand how these changes might influence arctic-nesting goose populations worldwide, it is critical to understand the population biology and ecology of these goose systems. It is therefore timely to examine these processes in the Greenland White-fronted Goose *Anser albifrons flavirostris*. This taxon is protected from hunting throughout almost its entire range, albeit illegal hunting is believed to persist at low levels. Site protection measures have been enacted on breeding, staging and wintering sites. In 1989, five sites on summering areas in west Greenland were designated as Wetlands of International Importance ("Ramsar Sites") under the terms of the Ramsar Convention and subsequently, in 2013, the main Icelandic staging site at Hvanneyri was also listed by Icelandic Government as a Ramsar Site. The entire world population winters in Great Britain and Ireland, where protection and increased food availability due to intensive agriculture have resulted in increases in most goose populations in recent decades. Indeed, there are 14 Ramsar Sites in Great Britain and 11 in Ireland utilised by Greenland White-fronted Geese. Yet the Greenland White-fronted Goose population has declined by 47% over the past two decades (Fig. 1) for reasons that are not clear (Fox *et al.* 2016).

Tackling a conservation challenge using population and individual level techniques

In this paper, following suggestions by Green (1995) and Gibbons *et al.* (2011), we synthesise a series of recently published results on Greenland White-fronted Goose demography, to diagnose the demographic mechanisms underlying population change of the geese at their main wintering site (Wexford, Ireland) for improved inference of factors influencing the global population decline. The studies were aided by long-term capture-mark-recapture (CMR), population size and productivity (*i.e.* the proportion of juveniles) datasets of Greenland White-fronted Geese from Wexford, which permitted estimates of the birds' survival and breeding success. Critically, these data encompassed a period of population increase between the early 1980s and the late 1990s (Fig. 1), and a subsequent decrease. Further, the percentage juveniles during winter (a productivity metric) at Wexford steadily declined from the early 1980s to mid-2010s (Fig. 2). Using these data, we were therefore able to examine and compare demographics associated with each period, and not just during the population decline. Very few studies on this population to date have been conducted on breeding grounds in Greenland, because of the remoteness of the area and the dispersed, low density of individuals across the landscape (Fox & Stroud 1988, 2002). Instead, most previous research has been carried out at staging sites (Francis & Fox 1987; Fox *et al.* 1999; Nyegaard *et al.* 2001; Fox *et al.* 2002; Fox *et al.* 2012) or wintering areas (Ruttledge & Ogilvie 1979; Mayes 1991; Wilson *et al.* 1991; Warren *et al.* 1992; Fox 2003). Novel tracking devices therefore were used to quantify behaviours and movements during the breeding season in Greenland. In particular we examined whether reductions

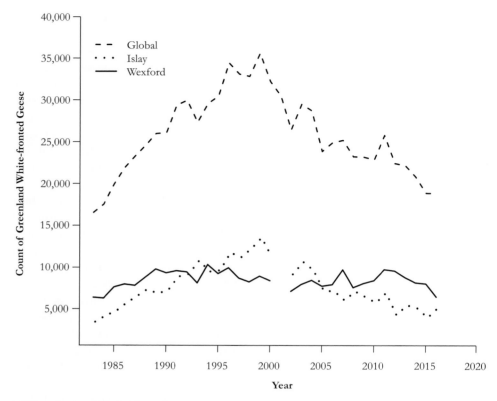

Figure 1. Greenland White-fronted Goose count of global population (dashed line), Islay subpopulation (*i.e.* wintering flock; dotted line) and Wexford subpopulation (solid line), 1983–2016.

in juvenile and adult survival, or in productivity, contributed to the population trajectories. This is a taxon with a complex social system and life history; previous studies have highlighted that these birds are long-lived (*e.g.* maximum age 22 years; A.D. Fox, unpubl. data) and exhibit prolonged family relationships (Warren *et al.* 1993). These factors may be interacting to influence population demography in subtle ways.

To examine the demographics of this system, we developed multistate models based on CMR life histories of collared birds, which estimated age- and sex-specific survival and movement probabilities. Our results suggested that there were no differences between sexes in emigration probabilities at ages 1 year (males: mean = 0.18, 95% credible intervals (CRI) = 0.14–0.22, females: mean = 0.17, 95% CRI = 0.13–0.22) and 2+ years (males: mean 0.11, 95% CRI = 0.09–0.14, females: mean = 0.11, 95% CRI = 0.08–0.13) or remigration probabilities (*i.e.* the return of birds to sites where they were originally marked after

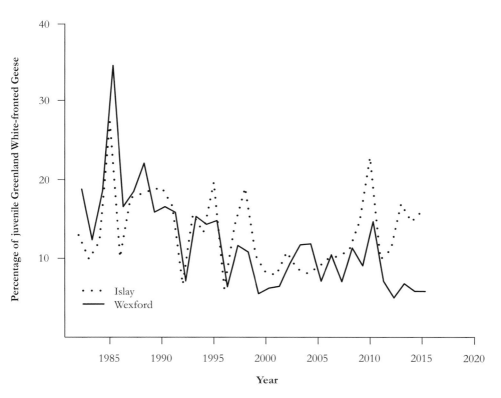

Figure 2. Percentage juvenile Greenland White-fronted Geese counted at Islay (dotted line) and Wexford (solid line), 1982–2015.

a period elsewhere) at ages 2+ year (males: mean = 0.11, 95% CRI = 0.08–0.15, females: mean= 0.13, 95% CRI = 0.09–0.18; Weegman *et al.* 2015). These findings confirmed published estimates of emigration (Marchi *et al.* 2010), and provided previously unknown estimates of remigration for this population.

This model framework formed the foundation for development of an integrated population model (IPM) which estimated age-, site- and year-specific survival and movement probabilities, and utilised population size and productivity

data to yield population growth rates over the 27-year study period (Weegman *et al.* 2016a). Importantly, observations of the Wexford population size showed that the flock has remained relatively stable during the study, despite large fluctuations (increases in the early period and declines in recent years) in the size of *c.* 70 other wintering flocks in Great Britain and Ireland. This includes a major increase and subsequent decline on Islay, Scotland, which is the second-largest wintering area (collection of flocks) for the subspecies, despite stable productivity for these birds

(Fig. 1, Fig. 2; Fox *et al.* 2016). Over the 27-year study period, the Wexford population constituted 25–42% of all Greenland White-fronted Geese globally and, prior to demographic analyses, appeared to be a classic source-sink system (Pulliam 1988), whereby large wintering aggregations (such as at Wexford) act as "sources" to support the smaller "sink" flocks, which explains persistence of the latter. However, our IPM revealed the exact opposite in that the largest concentration of wintering Greenland White-fronted Geese in the world (Wexford) is in fact a large sink, whose population size is maintained only by substantial annual immigration from other (smaller) sites (Weegman *et al.* 2016a). Indeed, model estimates of population growth rate reached *c.* 1.0 (the level required to match the observed stability at Wexford) only on invoking immigration at a remarkable *c.* 17% *per annum*.

Recruitment rate (*i.e.* a demographic measure of productivity) at Wexford generally declined over the study period, reflecting the observed percentage of juveniles there (Fig. 2). Taking into account juvenile and adult survival at Wexford, and based on previous approaches to identifying the causes of population declines through modelling exercises (Thomson *et al.* 1997; Robinson *et al.* 2004; Freeman *et al.* 2007), we identified that a reduction in productivity is an important demographic mechanism of Greenland White-fronted Goose population change at Wexford, but that this is masked by immigration (Fig. 3). Assuming factors contributing to Wexford's sink status occur during winter (and not where Wexford birds breed in Greenland or stage in Iceland),

these findings indicate that researchers should not necessarily use Wexford as a model of "optimal" environmental conditions (because the site functions as a sink). Researchers do however need to understand more about the constraints and restraints of this system throughout the annual cycle, including as it pertains to Wexford, and specifically whether successful breeding birds are leaving (thus creating the impression of low productivity there), the type of birds that move into Wexford (age, family status, reproductive success, *etc.*), and why they do so.

Worsening environmental conditions in Greenland explain cohort effects

In a population characterised by learned behaviour and complex social interactions (Fox 2003), declining productivity may be a product of subtle changes. For example, if adverse environmental conditions mitigate in favour of prolonged parent-offspring relationships, age at first successful reproduction would increase, causing a decline in *per capita* productivity. The environmental drivers of demographic change can be studied through cohort effects (Lindström 1999), because prevailing environmental conditions experienced by members of a hatch-year cohort affect their individual (and collective cohort) fitness, with subsequent impacts on population dynamics. On considering breeding success among cohorts (with successful reproduction measured as marked individuals returning to the wintering areas with young), environmental conditions (using North Atlantic Oscillation (NAO) values as a proxy) were found to

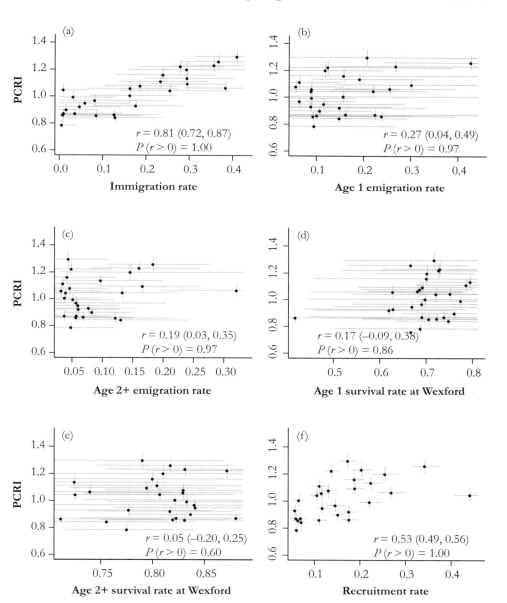

Figure 3. Immigration (a), emigration of geese aged 1 (b) and 2+ years (c), survival of geese aged 1 (d) and 2+ years (e) and recruitment rate (f) against *per capita* rate of increase per annum in population size (PCRI) of Greenland White-fronted Geese at Wexford, 1983–2010. Black dots show posterior means (with 95% CRI, grey lines). The posterior mode of the correlation coefficients (*r* with 95% CRI) and probability of a positive correlation (*P* (*r* > 0)) are inset (from Weegman *et al.* 2016a).

explain variation in age at first successful reproduction, but not the first successful brood size, for Greenland White-fronted Geese (Weegman *et al.* 2016b). Positive NAO values tended to result in relatively cold, dry conditions in west Greenland and were associated with earlier ages at first successful reproduction (the so-called "silver spoon" effect from Grafen 1988) and *vice versa*. However, the silver spoon effect was relatively minor and environmental conditions experienced in adulthood prior to successful reproduction and in the year of first successful reproduction had a much stronger influence on age at first successful reproduction. Cohorts bred successfully at younger ages when they experienced favourable environmental conditions, but this effect was far stronger when the preceding years were also favourable. These results confirm an association between environmental conditions on the breeding grounds and population productivity, and imply that such effects may be carried over multiple years.

The fitness implications of extended parent-offspring relationships

We also studied whether there was an association between the social system and productivity in Greenland White-fronted Geese. Previous work on this population suggested that some birds exhibit uniquely prolonged parent-offspring relationships (up to 6 years; Warren *et al.* 1993). However, the fitness implications of these extended relationships had never been studied. Our more recent analyses suggested that parent-offspring and sibling-sibling associations varied from 1–13 years but were only

beneficial through to age 3 years in Greenland White-fronted Geese, whereby fitness (*i.e.* survival and eventual breeding probability) of birds that maintained such associations was significantly greater than those that did not (Weegman *et al.* 2016c). Conversely, birds that maintained extended family associations (> 3 years) gained no significant fitness benefit over individuals that left parents or siblings at the same age. We combined these results to form a cost-benefit model, which suggested that fitness was lower among birds that remained with their parents or siblings than simulated birds who were forced into independence at ages 6 and 7 years. Although subsequent breeding probability was greatest for "older" individuals (*i.e.* those aged 5 years) associating with siblings, these gains were offset by non-significant survival differences between birds with siblings and those that were independent, yielding lower overall fitness for birds aged 5 years than those aged 3 (Weegman *et al.* 2016c). Independence after just 2 or 3 years may be especially important for species characterised by very few breeders or poor productivity such as Greenland White-fronted Geese because younger individuals have more potential breeding opportunities.

Other factors, such as density dependence, might limit productivity in this population. We studied whether these regulatory processes, which are known to occur in other systems (Newton 1998; Rodenhouse *et al.* 2003; Norris *et al.* 2004), influenced Greenland White-fronted Geese at the population and individual levels. Using IPM, we found for the Wexford population a strong positive correlation between emigration and immigration rates, and between emigration and recruitment rates, providing evidence of density-dependent regulation during winter. Hence, more birds immigrated to Wexford in years when more birds emigrated from Wexford; likewise more birds emigrated from Wexford in years when the recruitment rate at Wexford was larger. In our study of cohort effects, population sizes in the winter prior to hatch year or breeding year did not explain variation in age at first successful reproduction or the proportion of successful breeders by cohort (Weegman *et al.* 2016b). Thus, we found no evidence of density-dependent regulation of productivity. Overall, although the IPM suggested density-dependent processes may be occurring in this population, further studies are needed to: (i) examine the effect of Wexford regulation on the dynamics of the overall system, and (ii) determine whether similar density-dependent effects occur among other Greenland White-fronted Goose flocks.

Quantifying drivers of poor productivity at the individual level

Whilst we assessed the influence of environmental factors on fitness by linking age at first successful reproduction with NAO data, there are specific facets of breeding biology that we were not able to examine using these methods, but which also might explain the recent decline in productivity in Greenland. These might include decreased breeding propensity (the probability that an adult female attempts to breed in a given year), decreased clutch size, decreased incubation success, decreased

fledging success, or even lower survival during the post-fledging phase prior to winter (before juveniles are counted). These questions are difficult to answer because of the remoteness of the breeding range and low nesting densities of Greenland White-fronted Geese in west Greenland (Salomonsen 1967; Fox & Stroud 1988; Malecki *et al.* 2000). A novel method to answer these questions uses hybrid Global Positioning System (GPS) and acceleration (ACC) tracking devices to determine time- and energy-budgets throughout the year. We deployed these units on male geese during winter and downloaded data when the birds returned the following winter. Two of 15 tagged geese were categorised as having bred successfully during the study year because they were resighted repeatedly (> 5 times) with juveniles during winter. Thus, we used these individuals to understand the behavioural "traces" associated with the breeding event and compared these to birds resighted without juveniles during winter to determine whether individuals in the latter group deferred breeding or failed during incubation or chick-rearing. There were indications that the behaviour and energetics of birds that were either unsuccessful or deferred reproductive attempts diverged from those of the successful breeders early in the breeding season (Weegman 2014). However, with only two successful breeding birds tracked, larger sample sizes are needed for a robust assessment of whether low breeding propensity or high failure rates during early incubation are the most likely cause of low productivity in Greenland White-fronted Geese.

Constraints on breeding may occur in the pre-breeding migration and immediate post-migration periods, when it is assumed that individuals "prepare" for the anticipated cost of breeding by accumulating greater energy stores (Fox & Madsen 1981). To assess this, we compared: (a) the proportion of time feeding, and (b) energy expenditure, for tagged birds that bred successfully *versus* non-breeding (or failed-breeding) birds during spring migration and the pre-breeding period in Greenland (*i.e.* the first 14 days after arrival there), and found no substantive differences between these birds (Weegman 2014). That time- and energy-budgets through spring and early summer were indistinguishable between successful breeding birds and those without young in winter (*i.e.* which either failed in their nesting attempt or deferred breeding) suggests that all geese were "prepared" for a breeding attempt, and that any decision to defer was made immediately prior to incubation in Greenland, presumably in relation to the conditions encountered on breeding areas (Weegman *et al.* 2016b). It is possible, but we believe unlikely, that birds which decided not to attempt to breed made this decision before arrival in Greenland, and then simply showed no difference in time- and energy-budgets during spring and early summer.

These preliminary findings also suggest that Greenland White-fronted Geese not successful in reproduction are not limited by the amount of time spent feeding during spring migration from wintering to breeding areas (when birds must replenish energy stores). We infer this based on the observation that there was no difference in the proportion of time spent feeding

between successful breeders and non-breeders or failed breeders. Hence, these limited data do not provide support for carry-over effects in which condition during winter or spring influences productivity on breeding areas (Harrison *et al.* 2013), but are consistent with previous findings from field scores of abdominal profiles on wintering and staging areas during spring, which indicated that birds attained departure body condition on wintering and staging areas earlier in recent years than in any other period of the 27-year study (Fox *et al.* 2014), due to improved food availability (Francis & Fox 1987; Fox *et al.* 2012) and warmer springs (Fox *et al.* 2014). These shifts in phenology have increased the spring Icelandic staging period to one much longer than the birds require to replenish energy stores (Fox *et al.* 2014). Despite the low sample size, our findings build on previous Greenland White-fronted Goose work to suggest that the decision to lay a clutch is made on arrival in the breeding range. For instance, irrespective of whether the birds defer or fail in their breeding attempt, the lack of evidence to date for a carry-over effects supports the hypothesis of Boyd & Fox (2008) that increased spring snowfall in west Greenland in recent years (likely driven by changes in the NAO due to a warming climate; Hoerling *et al.* 2001; Johannessen *et al.* 2004) has created a phenological mismatch between weather conditions in Greenland and the timing of the breeding season. Hence, birds continue to arrive in west Greenland within a few days of historical arrival dates (Salomonsen 1950, 1967; Fox *et al.* 2014), but increased snow cover in some years may have reduced foraging opportunities. As a result, birds may be constrained to wait for snowmelt in order to reacquire necessary fat stores for reproduction. Phenological mismatches in chick-rearing and peak food abundance have been documented in a variety of other arctic-nesting birds, including Greater Snow Geese *Anser caerulescens atlantica* (Dickey *et al.* 2008) and Thick-billed Murres *Uria lomvia* (Gaston *et al.* 2009). However, further research on the ecosystem consequences of a warming arctic climate is necessary to understand how these processes potentially interact to affect the breeding biology of Greenland White-fronted Geese.

It is also possible that birds wintering in different parts of the range in Great Britain and Ireland have different time- and energy-budgets (especially energy expenditure) because of shorter or longer spring migration journeys. Greenland White-fronted Geese exhibit a "leap-frog" migration, whereby birds wintering in Scotland stage in the southern lowlands of Iceland and breed in the southernmost part of the breeding range in Greenland and those wintering in Ireland stage in western Iceland and breed in the northernmost part of the breeding range in Greenland (Salomonsen 1950; Kampp *et al.* 1988). These differences may lower productivity if birds migrating further do not feed more in advance of each stage of migration, because the greater energy expenditure associated with migrating longer distances would result in greater depletion of fat stores for these individuals than for those making shorter flights (assuming that northern breeding birds do not replenish nutrient stores in southern breeding areas before continuing

to their northern nest sites). We compared time- and energy-budgets using GPS-ACC data from birds wintering at Wexford, Ireland and Loch Ken, Scotland (Weegman *et al.* 2017). Although Irish birds flew significantly further than Scottish birds (but did not expend significantly more energy doing so), there were no significant differences in their proportion of time spent feeding during spring migration (*i.e.* from wintering to staging sites, staging to breeding sites and overall). These findings suggest plasticity in this species, whereby similar energy stores accrued by Irish and Scottish birds allow greater migration distances (of up to *c.* 300 km), if necessary. Nonetheless, that Scottish birds migrated significantly shorter distances suggests that they arrived in west Greenland with greater energy stores than Irish birds (assuming that Scottish birds were heavier than Irish birds on arrival in Greenland), which may facilitate greater reproductive success, as in other species (Ankney & MacInnes 1978; Newton 2008). Additional GPS-ACC data are needed to understand fully the extent to which migration distance influences reproductive success in Greenland White-fronted Geese, both directly on arrival at the breeding areas and as carry-over effects from wintering or staging areas.

One limitation of this study is that the majority of the long-term data we used was derived from one wintering site (Wexford). Although population survey data exist for *c.* 70 other wintering flocks, no consistent marking efforts have been conducted at these sites. It is now clear that there is a need to understand the dynamics of other flocks, particularly in the context of Wexford's function as a sink. The dynamics at the second-largest wintering area (Islay) are of greatest importance to understand in the short-term. Over our 27-year study period, the Islay population increased and subsequently decreased, mirroring global population fluctuations (Fig. 1). It would be particularly useful to understand what makes Wexford "more appealing" for immigrants than Islay and why apparently more families occur amongst birds wintering on Islay. One hypothesis might be that Islay does not function as a reserve like Wexford, where croplands and grasslands are managed for Greenland White-fronted Geese. In fact, an increasing Barnacle Goose *Branta leucopsis* population on Islay in recent years has led to shooting of those birds under license to decrease crop damage. It is possible that Greenland White-fronted Geese are experiencing increased disturbance resulting from shooting activities, perhaps discouraging immigration and encouraging emigration there. Results exploring disturbance on Islay are forthcoming (E. Burrell, unpubl. data). Furthermore, unlike Wexford, the population on Islay does not function as a single unit (*i.e.* one that roosts at one site) because population surveys and telemetry suggest there are over 50 separate roosts associated with different feeding areas on the island. Recent telemetry data suggests these flocks remain separated throughout winter. Thus, it might be more informative to study the dynamics of these flocks in the context of a geographical area with a particularly high density of small peripheral populations. To better understand the network of flocks at Islay and their relationship to the Wexford flock, consistent CMR efforts are necessary to model site demography in the IPM framework.

Diagnosis of population decline and future work

Using novel Bayesian IPMs, our diagnosis is that declining productivity measured at Wexford is the demographic mechanism for Greenland White-fronted Goose population change there, but immigration has masked these effects (Fig. 3; Weegman *et al.* 2016a). Declining productivity is possibly due to a reduced frequency of successful breeding, which could be the result of low breeding propensity or high failure rates in early incubation. Our work supports the suggestion that the drivers of low breeding success are likely to occur on breeding areas in west Greenland, perhaps related to weather conditions rather than carry-over effects from the preceding winter or spring. A priority for future work is to understand the processes occurring during the first 14 days after birds arrive in west Greenland, when foraging is required for replenishing energy stores prior to nesting (Fox & Madsen 1981). It is critical to understand whether increased snow cover is limiting forage availability and hence, given habitat requirements, the number of potential breeding territories (Fox & Stroud 1988). Additional conservation and management of breeding areas to increase productivity will be difficult because of the remoteness of such sites and the extent to which uncontrollable factors such as weather explain variation in productivity. Nonetheless, modelling exercises that estimate the survival rates required to match current low productivity rates would provide insight into whether additional spring and winter conservation efforts could mitigate poor productivity, assuming that survival can be increased further during these periods. Environmental variability in Greenland could also be included in such models in the IPM framework to predict population responses in future years. Indeed, if a warming climate is causing increased snowfall, further increases in temperatures will lead to rainfall (not snowfall), which might allow geese to arrive earlier, extend the breeding period and potentially improve habitat availability (Boyd & Fox 2008), as is currently the case for arctic-nesting geese in Svalbard (Jensen *et al.* 2008). Finally, these data could be incorporated into integrated metapopulation models to understand how processes at Wexford contribute to demography of the global population, to explain the consistent overall decline in recent years.

Acknowledgements

We thank the National Parks and Wildlife Service of Ireland, particularly the offices of John Wilson, David Norriss, David Tierney and the late Oscar Merne for their support. We also thank the many volunteers who have helped catch and mark Greenland White-fronted Geese at Wexford over the study period, especially Richard Hesketh, Carl Mitchell, Pat O'Sullivan, John Skilling, Arthur Thirlwell, and the late Chris Wilson. We thank Ran Nathan, Yehezkel Resheff and the Nathan lab for their assistance with analyses of GPS-ACC data. David Stroud and an anonymous referee made constructive comments on an earlier draft of the paper. Finally, we thank our respective employers for their support of this research. This research was funded through a joint Ph.D. studentship to M.D.W. from the Wildfowl & Wetlands Trust and the University of Exeter.

References

Abraham, K.F. & Jefferies, R.L. 1997. High goose populations: causes, impacts and implications. *In* B.D.J. Batt (ed.), *Arctic ecosystems in peril: report of the arctic goose habitat working group*, pp. 7–22. Arctic Goose Joint Venture Special Publication, U.S. Fish and Wildlife Service, Washington D.C., USA and Canadian Wildlife Service, Ottawa, Canada.

Ankney, C.D. & MacInnes, C.D. 1978. Nutrient reserves and reproductive performance of female Lesser Snow Geese. *The Auk* 95: 459–471.

Bantle, J.L. & Alisauskas, R.T. 1998. Spatial and temporal patterns in Arctic Fox diets at a large goose colony. *Arctic* 51: 231–236.

Boyd, H. & Fox, A.D. 2008. Effects of climate change on the breeding success of White-fronted Geese *Anser albifrons flavirostris* in west Greenland. *Wildfowl* 58: 55–70.

Cranswick, P.A., Raducescu, L., Hilton, G.M. & Petkov, N. 2012. *International Single Species Action Plan for the conservation of the Red-breasted Goose (*Branta ruficollis*).* AEWA Technical Series No. 46, The Wildfowl & Wetlands Trust, Slimbridge, UK.

Dickey, M.-H., Gauthier, G. & Cadieux, M.-C. 2008. Climatic effects on the breeding phenology and reproductive success of an arctic-nesting goose species. *Global Change Biology* 14: 1973–1985.

Durant, J.M., Hjermann, D.O., Ottersen, G. & Stenseth, N.C. 2007. Climate and the match or mismatch between predator requirements and resource availability. *Climate Research* 33: 271–283.

Fox, A.D. 2003. *The Greenland White-fronted Goose* Anser albifrons flavirostris. Doctor's dissertation (D.Sc.), National Environmental Research Institute, Denmark.

Fox, A.D. & Abraham, K.F. 2017. Why geese benefit from the transition from natural vegetation to agriculture. *Ambio* 46: 188–197.

Fox, A.D. & Madsen, J. 1981. The pre-nesting behaviour of the Greenland White-fronted Goose. *Wildfowl* 32: 48–53.

Fox, A.D. & Stroud, D.A. 1988. The breeding biology of the Greenland White-fronted Goose (*Anser albifrons flavirostris*). *Meddelelser Om Grønland, Bioscience* 27: 1–14.

Fox, A.D. & Stroud, D.A. 2002. Greenland White-fronted Goose *Anser albifrons flavirostris*. *Birds of the Western Palearctic Update* 4: 65–88.

Fox, A.D., Hilmarsson, J.Ó., Einarsson, Ó., Boyd, H., Kristiansen, J.N., Stroud, D.A., Walsh, A.J., Warren, S.M., Mitchell, C., Francis, I.S. & Nygaard, T. 1999. Phenology and distribution of Greenland White-fronted Geese *Anser albifrons flavirostris* staging in Iceland. *Wildfowl* 50: 29–43.

Fox, A.D., Hilmarsson, J.Ó., Einarsson, Ó., Walsh, A.J., Boyd, H. & Kristiansen, J.N. 2002. Staging site fidelity of Greenland White-fronted Geese in Iceland. *Bird Study* 49: 42–49.

Fox, A.D., Boyd, H., Walsh, A.J., Stroud, D.A., Nyeland, J. & Cromie, R.L. 2012. Earlier spring staging in Iceland amongst Greenland White-fronted Geese *Anser albifrons flavirostris* achieved without cost to refuelling rates. *Hydrobiologia* 697: 103–110.

Fox, A.D., Weegman, M.D., Bearhop, S., Hilton, G.M., Griffin, L., Stroud, D.A. & Walsh, A.J. 2014. Climate change and contrasting plasticity in timing of a two-step migration episode of an arctic-nesting avian herbivore. *Current Zoology* 60: 233–242.

Fox, A.D., Francis, I., Norriss, D. & Walsh, A. 2016. *Report of the 2015/16 International Census of Greenland White-fronted Geese*. Greenland White-fronted Goose Study, Rønde, Denmark.

Francis, I.S. & Fox, A.D. 1987. Spring migration of Greenland White-fronted Geese through Iceland. *Wildfowl* 38: 7–12.

Freeman, S.N., Robinson, R.A., Clark, J.A., Griffin, B.M. & Adams, S.Y. 2007. Changing demography and population decline in the Common Starling *Sturnus vulgaris*: a multisite approach to Integrated Population Monitoring. *Ibis* 149: 587–596.

Gaston, A.J., Gilchrist, H.G., Mallory, M.L. & Smith, P.A. 2009. Changes in seasonal events, peak food availability, and consequent breeding adjustment in a marine bird: a case of progressive mismatching. *The Condor* 111: 111–119.

Gauthier, G., Bêty, J., Giroux, J.-F. & Rochefort, L. 2004. Trophic interactions in a high arctic Snow Goose colony. *Integrative and Comparative Biology*, 44: 119–129.

Gauthier, G., Giroux, J.-F., Reed, A., Béchet, A. & Bélanger, L. 2005. Interactions between land use, habitat use, and population increase in Greater Snow Geese: what are the consequences for natural wetlands? *Global Change Biology* 11: 856–868.

Gibbons, D.W., Wilson, J.D. & Green, R.E. 2011. Using conservation science to solve conservation problems. *Journal of Applied Ecology* 48: 505–508.

Gilg, O., Kovacs, K.M., Aars, J., Fort, J., Gauthier, G., Grémillet, D., Ims, R.A., Meltofte, H., Moreau, J., Post, E., Schmidt, N.M., Yannic, G. & Bollache, L. 2012. Climate change and the ecology and evolution of arctic vertebrates. *Annals of the New York Academy of Sciences* 1249: 166–190.

Grafen, A. 1988. On the uses of data on lifetime reproductive success. *In* T.H. Clutton-Brock (ed.), *Reproductive Success: Studies of Individual Variation in Contrasting Breeding Systems*, pp. 454–471. University of Chicago Press, Chicago, USA.

Green, R.E. 1995. Diagnosing causes of bird population declines. *Ibis* 137 (Suppl.): 47–55.

Harrison, X.A., Hodgson, D.J., Inger, R., Colhoun, K., Gudmundsson, G.A., McElwaine, G., Tregenza, T. & Bearhop, S. 2013. Environmental conditions during breeding modify the strength of mass-dependent carry-over effects in a migratory bird. *PLOS ONE* 8: e77783.

Hoerling, M.P., Hurrell, J.W. & Xu, T. 2001. Tropical origins for recent North Atlantic climate change. *Science* 292: 90–92.

Intergovernmental Panel on Climate Change (IPCC). 2014. *Working Group II contribution to the Fifth Assessment Report of the Intergovernmental Panel on Climate Change.* IPCC, Geneva, Switzerland.

International Union for Conservation of Nature (IUCN) 2016. Red List of Threatened Species, version 2016.1. IUCN, Cambridge, UK. Available from www.iucnredlist.org (last accessed 5 October 2017).

Jensen, R.A., Madsen, J., O'Connell, M., Wisz, M.S., Tømmervik, H. & Mehlum, F. 2008. Prediction of the distribution of arctic-nesting Pink-footed Geese under a warmer climate scenario. *Global Change Biology* 14: 1–10.

Johannessen, O.M., Bengtsson, L., Miles, M.W., Kuzmina, S.I., Semenov, V.A., Alekseev, G.V., Nagurnyi, A.P., Zakharov, V.F., Bobylev, L.P., Pettersson, L.H., Hasselmann, K. & Cattle, H.P. 2004. Arctic climate change: observed and modeled temperature and sea-ice variability. *Tellus* 56A: 328–341.

Kampp, K., Fox, A.D. & Stroud, D.A. 1988. Mortality and movements of the Greenland White-fronted Goose *Anser albifrons flavirostris. Dansk Ornitologisk For Tidsskrift* 82: 25–36.

Lindström, J. 1999. Early development and fitness in birds and mammals. *Trends in Ecology and Evolution* 14: 343–348.

Madsen, J., Cracknell, G. & Fox, A.D., eds 1999. *Goose Populations of the Western Palearctic. A Review of Status and Distribution. Wetlands International Publication 48.* Wetlands International, Wageningen, the Netherlands and National Environmental Research Institute, Rønde, Denmark.

Malecki, R.A., Fox, A.D. & Batt, B.D.J. 2000. An aerial survey of nesting Greenland White-fronted and Canada Geese in west Greenland. *Wildfowl* 51: 49–58.

Marchi, C., Sanz, I.F., Blot, E., Hansen, J., Walsh, A.J., Frederiksen, M. & Fox, A.D. 2010. Between-winter emigration rates are linked to reproductive output in Greenland White-fronted Geese *Anser albifrons flavirostris. Ibis* 152: 410–413.

Mayes, E. 1991. The winter ecology of Greenland White-fronted Geese *Anser albifrons flavirostris* on semi-natural grassland and intensive farmland. *Ardea* 79: 295–304.

Newton, I. 1998. *Population Limitation in Birds.* Academic Press, London, UK.

Newton, I. 2008. *The Migration Ecology of Birds.* Academic Press, London, UK.

Nolet, B.A., Bauer, S., Feige, N., Kokorev, Y.I., Popov, I.Y. & Ebbinge, B.S. 2013. Faltering lemming cycles reduce productivity and population size of a migratory arctic goose species. *Journal of Animal Ecology* 82: 804–813.

Norris, D.R., Marra, P.P., Kyser, T.K., Sherry, T.W. & Ratcliffe, L.M. 2004. Tropical winter habitat limits reproductive success on the temperate breeding grounds in a migratory bird. *Proceedings of the Royal Society B: Biological Sciences* 271: 59–64.

Nyegaard, T., Kristiansen, J.N. & Fox, A.D. 2001. Activity budgets of Greenland White-fronted Geese *Anser albifrons flavirostris* spring staging on Icelandic hayfields. *Wildfowl* 52: 41–53.

Owen, M. 1990. The damage-conservation interface illustrated by geese. *Ibis* 132: 238–252.

Pulliam, H.R. 1988. Sources, sinks, and population regulation. *The American Naturalist* 132: 652–661.

Robinson, R.A., Green, R.E., Baillie, S.R., Peach, W.J. & Thomson, D.L. 2004. Demographic mechanisms of the population decline of the Song Thrush *Turdus philomelos* in Britain. *Journal of Animal Ecology* 73: 670–682.

Rodenhouse, N.L., Sillett, T.S., Doran, P.J. & Holmes, R.T. 2003. Multiple density-dependence mechanisms regulate a migratory bird population during the breeding season. *Proceedings of the Royal Society B: Biological Sciences* 270: 2105–2110.

Ruttledge, R.F. & Ogilvie, M.A. 1979. The past and current status of the Greenland White-fronted Goose in Ireland and Britain. *Irish Birds* 1: 293–363.

Salomonsen, F. 1950. *Grønland Fugle. The Birds of Greenland.* Munkgaard, Copenhagen, Denmark.

Salomonsen, F. 1967. *Fuglene på Grønland.* Rhodos, Copenhagen, Denmark.

Thomson, D.L., Baillie, S.R. & Peach, W.J. 1997. The demography and age-specific annual survival of Song Thrushes during periods of population stability and decline. *Journal of Animal Ecology* 66: 414–424.

Tulp, I. & Schekkerman, H. 2008. Has prey availability for arctic birds advanced with climate change? Hindcasting the abundance of tundra arthropods using weather and seasonal variation. *Arctic* 61: 48–60.

Wang, X., Fox, A.D., Cong, P., Barter, M. & Cao, L. 2012. Changes in the distribution and abundance of wintering Lesser White-fronted Geese *Anser erythropus* in eastern China. *Bird Conservation International* 22: 128–134.

Warren, S.M., Fox, A.D., Walsh, A. & O'Sullivan, P. 1992. Age at first pairing and breeding among Greenland White-fronted Geese. *Condor* 94: 791–793.

Warren, S.M., Fox, A.D., Walsh, A. & O'Sullivan, P. 1993. Extended parent-offspring relationships in Greenland White-fronted Geese (*Anser albifrons flavirostris*). *The Auk* 110: 145–148.

Weegman, M.D. 2014. The demography of the Greenland White-fronted Goose. Ph.D. thesis, University of Exeter, Cornwall Campus, Penryn, UK.

Weegman, M.D., Fox, A.D., Bearhop, S., Hilton, G.M., Walsh, A.J., Cleasby, I.R. & Hodgson, D.J. 2015. No evidence for sex bias in winter inter-site movements in an arctic-nesting goose population. *Ibis* 157: 401–405.

Weegman, M.D., Bearhop, S., Fox, A.D., Hilton, G.M., McDonald, J.L. & Hodgson, D.J. 2016a. Integrated population modelling reveals a perceived source to be a cryptic sink. *Journal of Animal Ecology* 85: 467–475.

Weegman, M.D., Bearhop, S., Hilton, G.M., Walsh, A. & Fox, A.D. 2016b. Conditions during adulthood affect cohort-specific reproductive success in an arctic-nesting goose population. *PeerJ* 4: e2044.

Weegman, M.D., Bearhop, S., Hilton, G.M., Walsh, A.J., Weegman, K.M., Hodgson, D.J. & Fox, A.D. 2016c. Should I stay or should I go? Fitness costs and benefits of prolonged parent-offspring and sibling-sibling associations in an arctic-nesting goose population. *Oecologia* 181: 809–817.

Weegman, M.D., Bearhop, S., Hilton, G.M., Walsh, A., Griffin, L., Resheff, Y.S., Nathan, R. & Fox, A.D. 2017. Using accelerometry to compare costs of extended migration in an arctic herbivore. *Current Zoology*, https://doi.org/10.1093/cz/zox056.

Wilson, H.J., Norriss, D.W., Walsh, A., Fox, A.D. & Stroud, D.A. 1991. Winter site fidelity in Greenland White-fronted Geese *Anser albifrons flavirostris*, implications for conservation and management. *Ardea* 79: 287–294.

Does organ and muscle plasticity vary by habitat or age in wintering Lesser Snow Geese *Anser caerulescens caerulescens*?

JÓN EINAR JÓNSSON[1]* & ALAN D. AFTON[2]

[1]University of Iceland, Research Centre at Snæfellsnes, Stykkishólmur, IS-340, Iceland.
[2]U.S. Geological Survey, Louisiana Cooperative Fish and Wildlife Research Unit, Louisiana State University, Baton Rouge, Louisiana 70803, USA.
*Correspondence author. E-mail: joneinar@hi.is

Abstract

Plasticity in organ and muscle size and function allows individuals to respond to changes in food quality or foraging behaviour, in accordance with cost-benefit hypotheses. Lesser Snow Geese *Anser caerulescens caerulescens* (hereafter Snow Geese) winter in rice-prairie and coastal-marsh habitats in southwest Louisiana, where the time that the birds spend foraging and walking, their composite diets, and associated fibre and energy contents, differ between these two habitats. We therefore hypothesised that: 1) Snow Geese that feed primarily in coastal marshes during winter would have larger digestive organs than those in rice-prairies, to adapt to the higher fibre content of their marsh vegetation diet; and 2) that leg muscles of Snow Geese feeding in rice-prairies would undergo greater hypertrophy and thus be larger than those in coastal marshes, because individuals in rice-prairie habitat spend more time walking while foraging. The first hypothesis applied to adults and juveniles alike, whereas under the second hypothesis, we knew from concurrent studies that juveniles walk more than adults and therefore predicted that they would have relatively larger leg muscles, after adjusting for body size. Seventy juvenile and 40 adult Snow Geese were dissected to test these two hypotheses about plasticity and hypertrophy with respect to habitat and foraging behaviour. Caeca and gizzard lengths were found to be larger for Snow Geese feeding in coastal marshes, where the food ingested is relatively high in fibre compared with the birds' diet in the rice-prairies. Conversely, leg muscles were larger for Snow Geese foraging in rice-prairies, where the juvenile geese spend relatively more time walking. Although not fully grown, juvenile Snow Geese also varied in the length of their digestive system and hypertrophy in muscles in relation to habitat, reinforcing the view that the birds' morphology adapts to different feeding habitats and diets.

Key words: diet, ecological segregation, geese, habitat selection, individual variation, Louisiana.

Wildfowl (2017) 67: 19–43

Plasticity (also termed morphological flexibility) in the length and function of digestive organs allows individuals to respond to variation in food quality or availability, or to respond to variation in foraging behaviour (Ankney & MacInnes 1978; Prop & Vulink 1992; Piersma & Lindström 1997; Starck 1999; Tielemann *et al.* 2003; Fox & Kahlert 2005; Williamson *et al.* 2014). Such changes can occur swiftly and repeatedly, and can also be reversible when needed. These seasonal changes represent trade-offs between the costs and benefits associated with maintaining organs or muscles at particular sizes, which can be explained in terms of prevailing ecological conditions and often expressed as the "cost-benefit hypothesis" (van Gils *et al.* 2003; Fox & Kahlert 2005). The adaptive value of these changes is to allow the organism to respond effectively to changes in ecological conditions (such as food types or food quality), so as to benefit individuals whilst the energetic costs of organ changes (*e.g.* carrying or maintaining larger or more complex organs) are kept to a minimum. Such evolutionary adaptations have been identified to meet specific or seasonal needs of breeding, migration and moult (Moorman *et al.* 1992; Prop & Vulink 1992; Piersma & Lindström 1997; Piersma *et al.* 1999; Starck 1999; van Gils *et al.* 2003; Fox & Kahlert 2005).

Plasticity is most commonly observed for the digestive system in birds, but also occurs for several other internal organs in response to seasonal changes, such as the heart and muscles (Ankney & MacInnes 1978; Piersma *et al.* 1999). Adaptations to changes in foraging conditions have been found to induce plasticity trade-offs in leg muscles, which can undergo atrophy or hypertrophy depending on how much running or walking is required within a season, or in areas with higher predation pressure (Fox & Kahlert 2005). Wild Barnacle Goose *Branta leucopsis* goslings become less active as the goslings mature and prepare to fly south and this decline in activity correlates with decreased aerobic capacity of the leg muscles (Bishop *et al.* 1998). However, it is unclear whether juvenile geese, which are not fully grown until after the first year of life (Davies *et al.* 1988; Cooch *et al.* 1991; Larsson & Forslund 1991), have developed organ systems capable of responding to environmental variation to the extent that adults are able to show organ plasticity. We nevertheless hypothesised that, although body size is not fully formed during the first winter of age in geese, the adaptive value of digestive organ plasticity in relation to habitat would provide benefit to both age groups (adults and juveniles).

Lesser Snow Geese *Anser caerulescens caerulescens* (hereafter Snow Geese) use rice-prairie and coastal-marsh habitats in southwest Louisiana during winter (Alisauskas *et al.* 1988; Alisauskas 1998; Jónsson & Afton 2006, 2015a, 2016; Jónsson *et al.* 2014). The relative costs and benefits to the geese in terms of food intake can vary annually for these two habitats in relation to weather or food availability (Alisauskas 1998; Alisauskas *et al.* 1988; Jónsson & Afton 2006; Jónsson *et al.* 2014). Coastal-marsh Snow Geese feed primarily on tubers of sedges *Scirpus* sp. (48% mean dry weight of vegetation consumed; Alisauskas *et al.* 1988) and on rhizomes of

Saltmeadow Cordgrass *Spartina patens* and Common Saltgrass *Distichlis spicata* (27% mean dry weight of plant vegetation consumed; Alisauskas *et al.* 1988). In contrast, rice-prairie Snow Geese ingest green vegetation, *i.e.* weedy forbs (70% mean dry weight of plant structures consumed) and graminoid leaves of rice plants (28% mean dry weight of plant structures consumed). Forbs comprise 13% crude fibre and 32% crude protein; tubers comprise 12% crude fibre and 7% crude protein; and rhizomes comprise 28% crude fibre and 4% crude protein (Alisauskas *et al.* 1988). Alisauskas *et al.* (1988) therefore defined the composite diets for Snow Geese wintering on the coastal marshes of southwest Louisiana as being of relatively high fibre content (20% dry weight) and low protein (8% dry weight; hereafter coastal-marsh diets). Corresponding values for rice-prairies were 15% and 27% dry weight for fibre and protein contents, respectively (hereafter rice-prairie diets). Digestibility of food, defined as the percentage of a given nutrient taken into the digestive tract that is absorbed into the body, generally is inversely related to fibre content and positively related to protein content (Sedinger & Raveling 1988; Prop & Vulink 1992; Sedinger 1997), and thus the marsh composite diet is less digestible than the corresponding rice-prairie composite diet.

Phenotypic changes to the gizzard or other digestive organs may be advantageous for: 1) improving the ability to grind plant materials that are rich in fibre, low in protein and therefore low in digestibility (Prop & Vulink 1992), such as *Scirpus* tubers in coastal marshes (Alisauskas 1988; Alisauskas

et al. 1988); or 2) obtaining a higher energy yield from the more fibrous diet, rather than simple physical effects of fibre content on food intake rates or on mechanical (muscular) compensation to fibrous food material (Williamson *et al.* 2014). A longer gut will prolong gut retention time, allowing increased absorption from diets with low digestibility (Prop & Vulink 1992). Accordingly, we hypothesised that Snow Geese (adults and juveniles alike) collected in coastal marshes would have larger digestive organs than those collected in rice-prairies.

Our earlier analysis of the behaviour of Snow Geese in the same study area found that their time-budgets differed between the rice-prairie and coastal marsh habitats, but these differences also varied with the age group of the birds. Adult geese on the rice-prairie spent, on average, 40% of their time feeding and 4.2% in locomotion, compared with 52% and 2% for those in coastal marshes. In contrast, juveniles on rice-prairies spent, on average, 54% of their time feeding and 5.1% in locomotion, compared with 41% and 1% respectively for juveniles in the coastal marshes (Jónsson & Afton 2006). Swimming was included in the locomotion category, but was rarely observed in the marsh where geese walked across vegetation mats. Because Snow Geese mostly actively search whilst feeding, we therefore hypothesised that among juveniles, Snow Geese would walk more in rice-prairies (the habitat where juveniles have to feed and walk for longer periods) than in coastal marshes, and thereby attain larger leg muscles. Although the Jónsson & Afton (2006) time-budget data indicated that adults also walk more when in coastal marsh habitat, the difference was less marked than for juveniles, so we predicted that leg muscles in this study would be of similar size across habitats for the adult geese.

A total of 110 Snow Goose specimens were collected and dissected to test the two main hypotheses. Firstly, that the digestive organs would differ in length by habitat, with habitat plasticity being evident for both age categories (*i.e.* adults and juveniles), and secondly that habitat-induced mobility would affect the size of leg muscles in waterfowl dependent on age, in that it would be evident in juveniles but not in adult birds.

Methods

Study area

Our study area (10,764 km²) in southwest Louisiana was bordered by Sabine National Wildlife Refuge (29°53'N, 93°23'W) on the west; Lake Charles and Highway 383 on the northwest; Highway 190 on the north; Highway 387 and Interstate 10 on the northeast; Highway 35 on the east, and the Gulf Coast on the south (see map in Jónsson *et al.* 2014). The Intra-coastal Canal generally separates the coastal marsh from the rice-prairies in southwest Louisiana (Bateman *et al.* 1988). Coastal marshes are comprised of fresh, intermediate, brackish or saline wetlands, but fresh and intermediate wetlands are not used frequently by Snow Geese. The coastal brackish and saline wetlands of the coastal marshes are about 32 km from the rice-prairies, which also are used by Snow Geese (Bateman *et al.* 1988). Rice-prairies are

former tall-grass prairies that have been extensively cultivated, mostly for rice, but also as pastures for cattle (Alisauskas 1988; Alisauskas *et al.* 1988; Bateman *et al.* 1988).

As with previous studies of Snow Geese in this area (Alisauskas *et al.* 1988, 1998; Jónsson & Afton 2016), we acknowledge that the birds may move between habitats and that, rather than staying in rice-prairies or coastal marshes for prolonged time periods, geese included in the study could have moved between habitats prior to collection. Our concurrent banding study however demonstrated that movements between habitats occurred only occasionally (Jónsson *et al.* 2014), and we therefore remain confident that, overall, the two groups were segregated and consistently exposed to diets of different fibre contents for a sufficiently long period to warrant a comparison of digestive organs between habitats. Despite these occasional movements, variation in skeletal morphology indicative of habitat segregation, first reported by Alisauskas (1998), was also evident during our 2002–2004 study (Jónsson 2005).

Collection of Snow Goose specimens

We examined gut and muscle samples from 70 juvenile and 40 adult Snow Goose specimens collected from 20 November to 17 February in the winters of 2001/02, 2002/03 and 2003/04, using 0.22 rifles and 12 gauge shotguns. The samples were collected within a 13 week period over the three different winters; variation in measurements attributable to collection date (measured as the number of days from 20 November each year) therefore was assessed in the subsequent analyses. These comprised 70 geese (21 adult females, 15 adult males, 15 juvenile females and 19 juvenile males) from the rice-prairies of Sweet Lake (8–16 km north of the Cameron Prairie National Wildlife Refuge; 29°51'N, 93°13'W) or in the vicinity of the towns of Lake Arthur at Oak Island (30°00'N, 92°04'W) and Thornwell (30°10'N, 92°80'W), and 40 geese (2 adult females, 2 adult males, 21 juvenile females and 15 juvenile males) from coastal marshes at Rockefeller State Wildlife Refuge (29°40'N, 92°55'W).

Collected specimens were individually double-bagged and frozen, and subsequently stored in a walk-in freezer at Louisiana State University. Collected specimens were sexed *post mortem* by cloacal examination (see Hochbaum 1942) and age was confirmed by plumage colour as either juvenile (hatch-year) or adult (see Baldassarre 2014). The geese were collected under the U.S. Fish and Wildlife Service's scientific collection permit MB048372-0, the Louisiana Department of Wildlife and Fisheries' scientific collection permit LNHP-01-052, and Louisiana State University Agricultural Center Institutional Animal Care and Use Committee's (LSU AgCenter IACUC) protocol number A01-09.

Digestive organ measurements

All specimens were thawed and measured, weighed and dissected (see Jónsson & Afton 2016). Specifically, we opened the abdominal cavity on the left side of each specimen, and carefully pulled out the alimentary tract. The gizzard was excised and measured with calipers (± 0.1 mm). The lengths of the following segments were disentangled, straightened and measured with a ruler (± 1 mm): 1) the upper digestive tract from the tip of the bill to the entrance to the gizzard, including the oesophagus and proventriculus (this approach provided reliable and repeatable start and end points for this measurement); 2) gizzard length; 3) the small intestine from the exit of the gizzard to the caecum; 4) both caeca; and 5) the large intestine.

Leg muscle measurements

We measured and weighed the paired gastrocnemius and tibialis anterior muscles (hereafter collectively termed "leg muscles"), which were cut from their attachment sites on the tibio-tarsus of the leg. These muscles were selected because others are more difficult to remove intact, and use of intact muscles ensures that the measurements are repeatable. Muscles were weighed with a digital scale to ± 0.1 g and measured immediately after excision. The muscles were then laid on a flat surface for measurement of muscle diameter (the response variable for leg hypertrophy), which was measured once, at the widest point of each muscle with digital calipers (± 0.1 mm).

Standardising data for body size effects

On analysing variation in digestive organ measurements, we included as response variables in linear mixed models the length of the upper digestive tract, the gizzard, the small and large intestines, and also the averaged lengths of the two caeca. For leg measurements, we included the weights and diameters of the gastrocnemius and tibialis muscles (4 measurements in total) as response variables in the analyses.

Researchers typically standardise measurements to some value representative of body size, particularly for species where there is known to be variation in body size across individuals that may influence the hypotheses being tested in the study (Relya 2005). We had good reasons to expect that overall body size would influence the sizes of our response variables (*i.e.* the length of the digestive organs or leg muscle diameters) for the Snow Geese. We therefore initially explored variation in body size in relation to age and sex, because we knew *a priori* that: 1) males are larger than females (Cooch *et al.* 1991; Alisauskas 1998; Jónsson 2005); 2) adults are larger than juveniles (Cooch *et al.* 1991; Jónsson 2005); and 3) that habitat can be related to body size in at least some years (Alisauskas 1998). We used a principal components analysis (PCA) on nine morphological measurements (skeletal morphology) to index body size (Alisauskas 1998; Jónsson & Afton 2016), using one PCA for all age and sex groups, so that PC1 would include body size variation attributable to sex, age, habitat and individual variability. The first principle component (PC1) of these nine morphological measurements explained 60.9% of the overall variation in this dataset and, thus, is a useful index of body size. We also considered the approach of Meixell *et al.* (2016) who used a segregated PCA to remove body size variation due to sex and age, but the results did not differ between the two PCA approaches, perhaps because in both analyses the effects of the PC1 were apparent only for upper digestive tract and

(weakly) for gizzard length. Here, we therefore present findings using the one PCA for all age and sex groups approach.

When PC1 was compared among adults and juveniles of each sex, adult males were larger than females in both age groups, females were similar in body size regardless of age, and juvenile males were intermediate between adult males and the two female groups (Appendix 1). We therefore standardised the organ measurements for body size effects by regressing PC1 on each of the digestive organ and leg muscle measurements to obtain residual values for each measurement for each individual. We then added the residual value to the overall mean measurement to obtain standardised individual size-adjusted values (mean + residual) (see Ankney & Afton 1988; Afton & Ankney 1991).

In addition to the habitat and age explanatory variables, we considered the effects of sex because males are larger than females (Alisauskas 1998; Jónsson & Afton 2016), and sex effects are commonly considered in studies of organ plasticity (Fox & Kahlert 2005; Laursen & Møller 2016). Sex, age and habitat were considered explanatory fixed effects because we included all possible groups of each variable. Adjusting our response variables with PC1 (overall body size) may conceivably have removed some of the variation associated with sex and age, but we chose to retain age and sex in the analyses because PC1 represents 49–61% of overall variation in the morphological measurements of wintering Snow Geese (Alisauskas 1998; Jónsson *et al.* 2014; Jónsson & Afton 2016; this study), and some of the body size variation therefore may not by represented by PC1 but rather by PC2, PC3, *etc*. We considered age in particular to be important, because juvenile Snow Geese are not fully grown until after 1 year of age (Cooch *et al.* 1991) and their internal organs may not be fully developed during the first winter of life. Age therefore was included in the analysis despite sample sizes being low for adults in the coastal marshes (*n* = 4), albeit higher for adults on rice-prairies (*n* = 36), because the data were obtained initially for a study on foraging by juvenile Snow Geese (Jónsson & Afton 2016). Comparisons of adult and juvenile morphologies in relation to habitat are consequently interpreted with consideration to these sample size restrictions in the current study.

Sampling effort also was not evenly distributed across the three winters (2001/02–2003/04 inclusive), with 2002/03 accounting for 67% of our collected specimens (23 adults; 51 juveniles). It therefore was not feasible to include winter as an explanatory variable in the model, especially not in relation to sex or age. Given that most of the birds were collected in a single winter, and that within-winter effects may be more important than annual variation in influencing gut and muscle measurements (because of the length of time that the birds have been on the two habitats since the start of the winter), we considered temporal variation by including collection date as a random effect and assumed that effects of collection date were independent of winter. We know from our sample of neck-collared Snow Geese (Jónsson & Afton 2014) that their mean

(± s.d.) duration of stay in Louisiana was 57.3 (± 32.3) days (J.E. Jónsson & A. Afton, unpubl. data), and that collection date can serve as an index of how long each specimen had been in southwest Louisiana. We compared standardised regression coefficients (*i.e.* betas from z-scored variables) for meaningful correlations (indicated by AIC) between collection date and each measurement, to evaluate which measurement was mostly influenced by collection date.

Linear mixed model analysis

The body size adjusted measurements for each digestive organ were used as response variables in linear mixed model analysis. A second PCA was used to reduce the dimensions of the body size adjusted muscle measurements from four dimensions to one before running the linear mixed models. Here, the first PC score (hereafter termed leg muscles) explained 62.3% of the overall variation in the four muscle measurements.

AIC model selection (Anderson 2008) was used to compare linear mixed models for the PC scores between sex (male or female) and habitats (rice-prairies *versus* coastal marshes), using the AICcmodavg package in the R system (Mazerolle 2015). We followed Burnham and Anderson (2002) and Anderson (2008), in considering that models which make no biological sense need not be tested. We therefore included only interactions relating to the study hypotheses in our models, in particular the habitat * age interaction to test the hypothesis that organ plasticity by habitat is independent of age, although the low sample size for adults in coastal marshes requires careful interpretation of this interaction for this dataset. We did not include the habitat * sex interaction because we had no biological reasons to believe that sex effects would depend on habitat, or *vice versa* (*i.e.* that one sex would be affected by habitat differences and not the other). Most adult birds are paired, so the sexes generally do not segregate by habitat, and males and females therefore should respond to habitat variation independently of sex. The age * sex interaction was included because data inspection suggested that age effects were dependent on sex (*i.e.* adult males were larger in overall body size than juvenile males), whereas adult and juvenile females were similar in overall body size (Appendix 1).

Model building followed a four-step process: 1) we commenced with a model which included all of the fixed explanatory variables (Habitat + Sex + Age), and then added all nested models, including the single effects models for each of these variables; 2) collection date was included as a random effect in all models and calculation of AIC values were repeated; 3) the age * sex interaction was added to all models that included both age and sex, and the habitat * age interaction was added to all models that included both habitat and age; and lastly 4) we ran null models (models with intercept only) for all analyses, resulting in 26 models.

Following Anderson (2008), we identified pretender variables as follows: candidate models which are within ΔAIC $\leq$ 2.0 of the top-ranked model but differ from the top-ranked model by the inclusion of 1–2 additional variables, yet their log-likelihood values are almost the same as those of the top-ranked model (see also Jónsson & Afton 2016). Since we were running six model selections on six response variables, we also examined cumulative weights of variables to compare the relative importance of habitat, sex, age, and collection date relative to one another. Cumulative weights were calculated prior to running the last models, which contained the age * sex or habitat * age interactions, to ensure an even number of models for each variable.

Results

Generally, there were varying degrees of support for the explanatory variables among the six analyses (Tables 1a–1f), with only two analyses sharing the same top-ranked model, *i.e.* the model containing all five explanatory variables for small intestine and gizzard length. Effects of collection date and habitat or habitat * age were supported in four analyses and effects of sex * age were supported in three analyses. Of the correlations with collection date, caeca length had the highest beta (0.158) whereas small intestine, large intestine and gizzard length had similar betas of 0.021, 0.036, and 0.046, respectively.

The best supported model for upper digestive tract was sex + age + sex * age, whereas there was no support for other models except habitat + sex + age + sex * age (ΔAIC = 1.3) and sex + age (ΔAIC = 1.7, Table 1a). However, we identified habitat as a pretender variable because of the small change in LogL (0.4) between habitat + sex + age + sex * age   and   the top-ranked   model   sex + age + sex * age. Cumulative weights for sex and age were both 1.00, compared to 0.30 and 0.06 for

Table 1. The top ten of 26 linear mixed models testing effects of sex, habitat, age, sampling date and collection date on the length of the digestive organs and leg muscles from 110 Snow Geese, collected in southwest Louisiana in winters 2002/03–2003/4. Models used for interpretation are shown in bold, see text for details.

a) Length of the upper digestive tract

Rank	Model	K	AIC	ΔAIC	Wi	LogL
1	**Sex + Age + S * A**	**5**	**406.0**	**0.0**	**0.393**	**−197.7**
2	Habitat + Sex + Age + S * A	6	407.4	1.3	0.204	−197.3
3	Sex + Age	4	407.8	1.7	0.165	−199.7
4	Habitat + Sex + Age	5	409.4	3.4	0.073	−199.4
5	Habitat + Sex + Age + H * A + S * A	7	409.6	3.6	0.065	−197.3
6	Sex + Age + Collection date + S * A	6	410.9	4.8	0.035	−199.0
7	Habitat + Sex + Age + H * A	6	411.6	5.6	0.024	−199.4
8	Habitat + Sex + Age + Collection date + S * A	7	412.7	6.6	0.014	−198.8
9	Sex + Age + Collection date	5	413.3	7.2	0.011	−201.3
10	Habitat + Sex + Age + Coll. date + H * A + S * A	8	413.4	7.4	0.010	−198.0

(b) Length of caeca

Rank	Model	K	AIC	ΔAIC	Wi	LogL
1	**Habitat + Age + Collection date + H * A**	**6**	**544.2**	**0.0**	**0.213**	**−265.7**
2	Habitat + Sex + Age + Coll. date + H * A	7	545.9	1.7	0.092	−265.4
3	Habitat + Sex + Age + Coll. date + H * A + S * A	8	546.0	1.8	0.088	−264.3
4	Habitat	3	546.2	2.0	0.077	−270.0
5	Habitat + Collection date	4	546.5	2.3	0.068	−269.1
6	Habitat + Age + Collection date	5	546.5	2.3	0.067	−268.0
7	Habitat + Age	4	546.9	2.6	0.057	−269.2
8	Habitat + Sex + Age + H * A + S * A	7	547.0	2.8	0.053	−268.3
9	Habitat + Age + H * A	5	547.5	3.3	0.041	−268.5
10	Habitat + Sex + Collection date	5	548.0	3.7	0.033	−268.7

Table 1 (*continued*).

(c) Length of small intestines

Rank	Model	K	AIC	ΔAIC	Wi	LogL
1	**Habitat + Sex + Age + Coll. date + H * A + S * A**	**8**	**868.8**	**0.0**	**0.673**	**−425.7**
2	Habitat + Sex + Age + Coll. date + H * A	7	871.8	3.0	0.147	−428.3
3	Habitat + Sex + Age + Collection date + S * A	7	872.6	3.8	0.099	−428.7
4	Sex + Age + Collection date + S * A	6	875.4	6.7	0.024	−431.3
5	Habitat + Age + Collection date + H * A	6	875.5	6.8	0.023	−431.3
6	Habitat + Sex + Age + Collection date	6	875.7	6.9	0.021	−434.4
7	Sex + Age + Collection date	5	878.7	9.9	0.005	−434.0
8	Habitat + Age + Collection date	5	879.5	10.8	0.003	−433.1
9	Habitat + Sex + Collection date	5	880.1	11.3	0.002	−434.8
10	Sex + Collection date	4	882.2	13.4	0.001	−436.9

(d) Length of large intestines

Rank	Model	K	AIC	ΔAIC	Wi	LogL
1	**Collection date**	**3**	**393.3**	**0.0**	**0.226**	**−193.5**
2	Age + Collection date	4	393.6	0.3	0.192	−192.6
3	Habitat + Collection date	4	395.2	1.9	0.086	−193.4
4	Habitat + Age + Collection date	5	395.7	2.5	0.065	−192.6
5	Sex + Collection date	4	396.1	2.8	0.056	−193.8
6	Habitat + Age + Collection date + H * A	6	396.3	3.0	0.050	−191.7
7	Age	3	396.4	3.1	0.048	−195.1
8	Sex + Age + Collection date	5	396.6	3.3	0.044	−193.0
9	Habitat	3	397.1	3.9	0.033	−195.5
10	Habitat + Age	4	397.6	4.3	0.026	−194.6

Table 1 (*continued*).

(e) Gizzard lengths

Rank	Model	K	AIC	ΔAIC	Wi	LogL
1	**Habitat + Sex + Age + Coll. date + H * A + S * A**	**8**	**687.8**	**0.0**	**0.607**	**−335.2**
2	Habitat + Sex + Age + Collection date + S * A	7	690.1	2.3	0.188	−337.5
3	Habitat + Sex + Age + Collection date + H * A	7	691.4	3.6	0.098	−338.2
4	Habitat + Age + Collection date + H * A	6	693.6	5.8	0.033	−340.4
5	Habitat + Sex + Age + Collection date	6	693.8	6.0	0.030	−340.5
6	Habitat + Sex + Collection date	5	695.5	7.7	0.013	−342.5
7	Habitat + Age + Collection date	5	695.9	8.1	0.011	−342.6
8	Habitat + Collection date	4	697.2	9.5	0.005	−344.4
9	Habitat	3	698.2	10.5	0.003	−346.0
10	Habitat + Sex	4	698.4	10.6	0.003	−345.0

(f) Length of leg muscles

Rank	Model	K	AIC	ΔAIC	Wi	LogL
1	**Habitat + Age**	**4**	**359.6**	**0.0**	**0.433**	**−175.6**
2	Habitat + Sex + Age	5	361.5	1.9	0.165	−175.5
3	Habitat + Age + H * A	5	361.7	2.1	0.155	−175.5
4	Habitat + Sex + Age + S * A	6	363.6	4.0	0.059	−175.4
5	Habitat + Sex + Age + H * A	6	363.6	4.0	0.059	−175.4
6	Habitat + Age + Collection date	5	365.1	5.5	0.028	−177.3
7	Age	3	365.3	5.7	0.025	−179.5
8	Habitat + Sex + Age + H * A + S * A	7	365.8	6.1	0.020	−175.3
9	Habitat + Age + Collection date + H * A	6	366.1	6.5	0.017	−176.6
10	Sex + Age	4	366.9	7.3	0.011	−179.3

habitat and collection date, respectively. Adult males had longer upper digestive tracts than the other age and sex groups, there was more overlap between the age groups within females than males, and juvenile males were larger than juvenile females and similar to the larger 50% among adult females (Fig. 1).

The analysis for caeca supported effects of habitat * age and collection date (Table 1b). We identified sex and sex * age as pretender variables, with a cumulative weight of 0.29 for sex and small difference in LogL values (0.3 and 1.4) between the highest ranked model habitat * age + collection date and the corresponding models (ΔAIC = 1.7 and 1.8) that also included sex or sex * age. Specimens from coastal marshes had larger caeca than those from rice-prairies, and among rice-prairie specimens, caeca length was smaller in

adults than juveniles, while there was no age effect in coastal marshes (Fig. 2a). Caeca length increased with collection date (Fig. 2b).

The model containing all five explanatory variables was best supported for small intestine (Table 1c). However, cumulative weights were 0.99, 0.89, 0.87, and 0.80 for collection date, age, sex, and habitat, respectively, indicating a varying degree of support for of them. Small intestine length was positively correlated with collection date (Fig. 3a). Adults of both sexes were larger but also more variable than juveniles, whereas there was little size difference by sex in juveniles compared to that in adults (Fig. 3b). Small intestines from rice-prairies generally were more variable than those from coastal marshes, and adult coastal marsh specimens were larger than the other age and habitat groups (Fig. 3c).

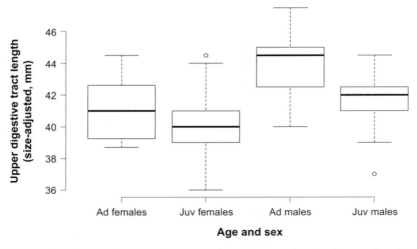

Figure 1. Differences between the sexes and age groups in length of upper digestive tracts (mm) for adult (*n* = 40) and juvenile (*n* = 70) Snow Geese collected in southwest Louisiana during winters 2001/02–2003/04. Tukey boxplots: the length of the box is the interquartile range; whiskers are drawn to the largest observations within 1.5 interquartile lengths from the top and bottom.

(a)

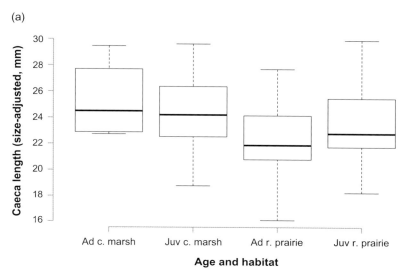

(b)

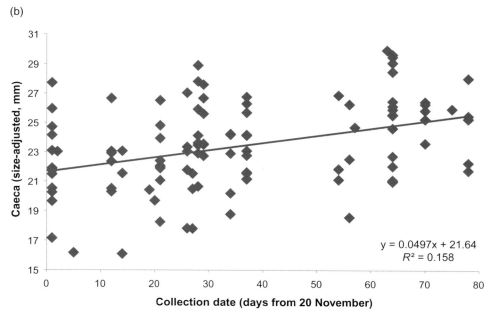

Figure 2. Differences in length of caeca (mm): a) between the habitats (coastal marshes and rice-prairies) and age groups, and b) by collection date, for adult ($n = 40$) and juvenile ($n = 70$) Snow Geese collected in southwest Louisiana during winters 2001/02–2003/04. Tukey boxplots: the length of the box is the interquartile range; whiskers are drawn to the largest observations within 1.5 interquartile lengths from the top and bottom.

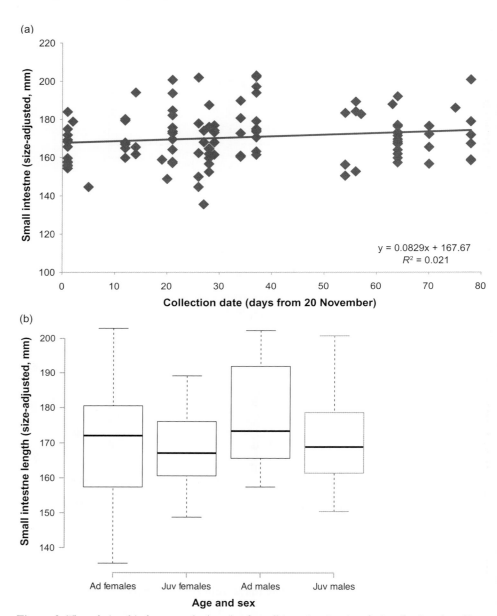

Figure 3. The relationship between the length of small intestine (mm) and: a) collection date, b) age and sex categories, and c) age groups and habitats (coastal marshes and rice-prairies), based on specimens of adult (*n* = 40) and juvenile (*n* = 70) Snow Geese collected in southwest Louisiana during winters 2001/02–2003/04. Tukey boxplots: the length of the box is the interquartile range; whiskers are drawn to the largest observations within 1.5 interquartile lengths from the top and bottom.

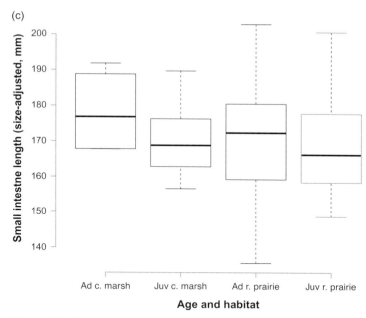

Figure 3 (*continued*).

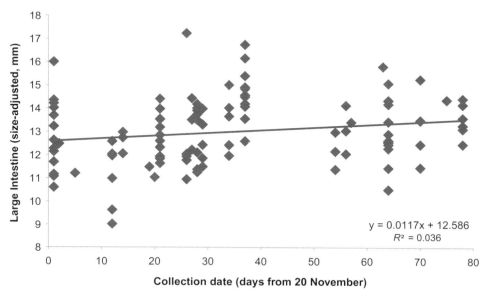

Figure 4. Relationship between collection date and the length of large intestine (mm) for adult (*n* = 40) and juvenile (*n* = 70) Snow Geese collected in southwest Louisiana during winters 2001/02–2003/04.

(a)

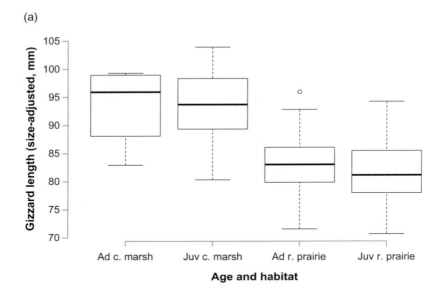

(b)

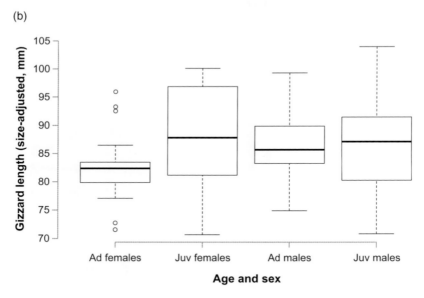

Figure 5. Differences in the length of gizzards (mm): a) between age groups and habitats (coastal marshes and rice-prairies); (b) between sexes and age groups, and c) in relation to collection date, for adult ($n = 40$) and juvenile ($n = 70$) Snow Geese collected in southwest Louisiana during winters 2001/02–2003/04. Tukey boxplots: the length of the box is the interquartile range; whiskers are drawn to the largest observations within 1.5 interquartile lengths from the top and bottom.

(c)

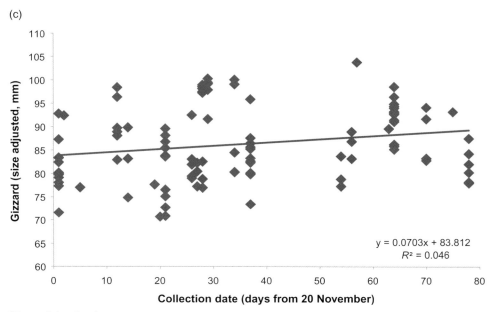

Figure 5 (*continued*).

For the large intestine, the highest-ranked models were collection date (single effects model) and age + collection date (ΔAIC = 0.3; Table 1d). However, the LogL difference between the two models was 0.9, suggesting that age was a pretender variable. Furthermore, collection date had the highest cumulative weight of 0.80, whereas age had cumulative weights of 0.47, indicating poor support for the effects of age. Similarly, the cumulative weights of habitat and sex indicated no support for those variables; 0.30 and 0.21, respectively. Length of the large intestine was positively correlated with collection date (Fig. 4).

The model containing all five explanatory variables was best supported for gizzard length (Table 1e) and the cumulative weights were 1.00, 0.85, 0.69, and 0.65 for habitat, collection date, sex and age, respectively.

Habitat was by far the best supported variable; all models containing habitat (ranks 1–16) had ΔAIC $\leq$ 13.6, whereas all models without habitat (ranks 17–26) had ΔAIC $\geq$ 46.7 (Table 1e). Specimens from coastal marshes had larger gizzards than those from rice-prairies (Fig. 5a). Gizzard length in coastal marshes was also more variable than that in rice-prairies and adults were slightly larger within both habitats, although our sample size may not have been sufficient to detect statistically an age effect within coastal marshes (Fig 5a). The sex * age interaction suggested that adult females had smaller gizzards than other age and sex groups, and that gizzard length was more variable within juveniles of both sexes than for adults (Fig. 5b). Collection date was well supported (cumulative weight 0.85) and present in the eight top-ranked models

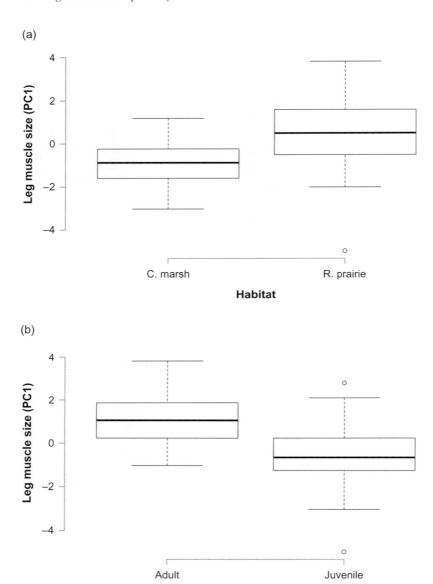

Figure 6. Differences in leg muscle size (PC1 from leg muscle measurements) between: a) habitats (coastal marshes and rice-prairies), and b) age groups, for adult (*n* = 40) and juvenile (*n* = 70) Snow Geese collected in southwest Louisiana during winters 2001/02–2003/04. Tukey boxplots: the length of the box is the interquartile range; whiskers are drawn to the largest observations within 1.5 interquartile lengths from the top and bottom.

(ΔAIC $\leq$ 8.1); gizzard length was positively correlated with collection date (Fig. 5c).

For leg muscles, the habitat + age model was the lone best supported model (Table 1f) and cumulative weights for habitat and age were 0.94 and 1.00, respectively. There was no support for effects of sex or collection date, where the cumulative weights were 0.27 and 0.06, respectively. Specimens from rice-prairies had larger leg muscles than those from coastal marshes (Fig. 6a) and adults had larger leg muscles than juveniles (Fig. 6b) but effects of habitat and age were independent of each other.

Discussion

The results of our study indicated that the digestive organ morphology of juvenile Snow Geese differed between habitats, as shown previously for adult Snow Geese (see also Alisauskas 1988; Alisauskas *et al.* 1988). Digestive organs were larger in coastal marshes, whereas leg muscles were larger in rice-prairies (*i.e.* the habitat that required more mobility), which was consistent with our hypothesis. However, contrary to our prediction, the habitat effect on leg muscles was independent of age. Caeca, small intestine, large intestine and gizzard length were all related to collection date. Of those, caeca seemed to have the strongest relationship to collection date, suggesting that caeca respond more strongly to the winter diet than the other digestive organs.

We found that coastal-marsh specimens had larger caeca and gizzards than did those from rice-prairies. This difference in the length of the gizzards seemingly is related to the differences in the protein and fibre contents and digestibility of the composite diets between habitats (Alisauskas *et al.* 1988). A larger gizzard in birds on a high-fibre diet can be interpreted as an adaptation towards increased digestibility of dietary fibre (Williamson *et al.* 2014). Digestive organs of waterfowl generally increase in size in response to increased fibre contents (Miller 1975; Paulus 1982; Halse 1984; Thompson & Drobney 1996), but such changes in Snow Geese seem more prominent in the caeca and gizzard than in the remainder of the digestive tract. Such a difference in gizzard length requires prolonged exposure to high-fibre diets in coastal marshes, as empirical data suggests that short-term exposure to high-fibre diets does not necessarily cause an increase in gizzard length (Jones *et al.* 2013).

The smaller caeca of adult specimens in rice-prairies, compared to that of juveniles in both habitats and adults in coastal marshes, were somewhat surprising but may have resulted from years of low fibre diets for these adult specimens, causing their caeca to become regressed from their first winter of life until the date they were collected. Alternatively, juveniles are inexperienced foragers and may be less adept than adults in maximising low fibre food when foraging, and therefore may develop larger caeca in response to the high fibre intake that results from their less effective selection of low fibre foods which adults are more adept at finding. Like other herbivorous waterfowl, geese have limited capacity to digest cellulose to a certain extent, via bacterial flora within the caeca. Generally, the overall contribution to total energy intake from cellulose digestion is small, mostly because of the fast throughput time of food (*c.* 70–120 min) through the alimentary canal (Mattocks 1971; Buchsbaum *et al.* 1986; Prop & Vulink 1992). Conversely, the caeca occasionally may function in some storage capacity until their contents are used to supplement the diet. This relative unimportance of cellulose for nutrition may explain why caeca in our coastal marsh specimens were not enlarged as much as the gizzard in response to a more fibre rich diet in the coastal marshes. Lastly, caeca are notably shorter in geese than in grouse (which also are predominantly herbivorous but often relatively non-migratory birds), which could be explained by energetic costs of carrying large caeca (or perhaps a cost of maintaining them via the circulatory or endocrine systems) during the long migrations undertaken by geese (Sedinger 1997).

The interactive terms habitat * age and sex * age did not provide additional explanatory information for variation in the length of the large intestine or leg muscles. Conversely, the habitat * age interaction was found to influence caeca length, sex * age influenced the upper digestive tract, and both interactions added information to the analyses of small intestine and gizzard lengths. Our data suggests that coastal marsh Snow Geese have larger gizzards and small intestines than those in rice-prairies, but these differences may not be independent of age in that: 1) adults in rice-prairies have slightly larger gizzards than juveniles (after adjusting for age differences in body size) but that gizzard length in coastal marshes is more variable and possibly independent of age; 2) adults have longer small intestines than juveniles but

either the greater variation among rice-prairie specimens may blur age effects within that habitat, or our small sample size among adults may do that in coastal marshes; 3) the sex * age effect on gizzard length may be a result of our sample being biased towards smaller gizzards in adults found in rice-prairies (where 90% of our adults were collected); and 4) our habitat * age findings could come from the analysis failing to find an age effect in coastal marshes but finding it in rice-prairies.

Waterfowl do not have a specialised crop or a comparable storage organ but the oesophagus is capable of expanding to accumulate substantial amounts of food, especially during short feeding bouts, allowing geese to store food to subsequently digest at roost sites (Owen 1980). Effects of age * sex were the only plausible explanatory variables for variation in the length of the upper digestive tract, which was larger in males within both age groups, but more interestingly, juvenile males had longer upper digestive tracts than did the majority of the adult females. We interpret these differences as an enhanced sexual dimorphism particular to neck length, which is not represented by the sex difference in overall body size (PC1), *i.e.* a sexually selected size difference in addition to differences overall body size. Male geese generally spend more time alert than females or juveniles (Gauthier & Tardif 1991; Fowler & Ely 1997; Flint *et al.* 1997). Thus, we hypothesise that males have evolved longer necks (and concurrently, longer upper digestive tracts) as a result of sexual selection. These benefits of higher posture and greater visibility, or improved ability to appear more threatening during aggressive displays have had value for reproductive success throughout the evolutionary history of Snow Geese. As a result of greater vigilance, males may feed less for long periods and also use shorter feeding bouts as a result of their time investment spent alert. Thus, there is another potential benefit of longer necks that relates to food ingestion or possibly digestion: the longer neck and upper digestive tract may allow males to ingest more food quickly when it is superabundant. This adaptation also could lengthen food throughput time by lengthening the digestive tract to enhance digestive capacity (*cf.* Prop & Vulink 1992). Somewhat surprisingly, this sexual dimorphism is already evident in juveniles but perhaps it is costly to grow the neck tissues, and thus, the process begins early and extends over a couple of years until a longer neck conveys a competitive advantage as the juvenile males become mature and obtain mates and subsequently families.

Gizzards of herbivorous birds often contain grit (gastroliths) which aid the breakdown of tough plant material (Williamsson *et al.* 2014). Snow Geese in the study area are heavily dependent on grit sites (patches of sand and gravel, commonly created by managers for gizzard-grit consumption by waterfowl), which are located within both rice-prairies and coastal marshes (Harris 1990). We did not analyse gizzard contents or attempt to measure grit in our Snow Goose samples because the bulk of these "gastroliths" was merely sand from these grit sites. Snow Geese in coastal marshes may consume more sand from grit sites than those in rice-prairies but there is no reason to expect different types of gastroliths between gizzards from the two habitats. Thus, gizzard length probably is not affected much by gastrolith types but rather the different composite diets, possibly their different energy contents (Alisauskas *et al.* 1988).

In conclusion, the analyses found that Snow Geese in coastal marshes had larger gizzards and caeca than those on rice-prairie habitat, which may be attributable to the fibre contents of the different composite diets (Alisauskas *et al.* 1988). Our findings for juveniles indicate that organ plasticity in relation to habitat begins during the first winter of life, despite the fact that juveniles have not attained full growth. This provides a useful baseline for exploring in further detail the effects of different habitat types on organ and muscle plasticity, as expressed by differential muscular exertion and different energy and fibre content of composite diets. These findings could also prove important in the face of the continued northward expansion of the Snow Goose wintering range (Jónsson & Afton 2015b), which could decrease the need for organ plasticity among adult or juvenile Snow Geese.

Acknowledgements

Our study was funded by the Canadian Wildlife Service, Louisiana Department of Wildlife and Fisheries (LDWF), Delta Waterfowl Foundation, Rockefeller Scholarship program, and the USGS-Louisiana Cooperative Fish and Wildlife Research Unit, Graduate School, Agricultural Center, and School of Renewable

Natural Resources at Louisiana State University. We are very grateful to D.G. Homberger for discussions about muscle measurements. We thank Eileen Rees, Tony Fox, and especially two anonymous reviewers for helpful comments on the final manuscript. We also thank D. Blouin, B. Barbe, M. Chamberlain, W. Henk, R.N. Helm (deceased), D. Caswell, R. Elsey, G. Perrie, C. Jeske (deceased), M. Kaller, J. Linscombe, T. Hess (deceased), M. Hoff, W. Norling, and S. Lariviere for their assistance, valuable input and support. We especially thank the staff of Rockefeller SWR, Cameron Prairie NWR, Sabine NWR, Sweet Lake Land and Oil Company and LDWF, who kindly provided housing and logistical support. C.J. Michie, B. Meixell, M. Pollock, T. Blair, J. Yurek, S. Kinney, and staff of the Sweet Lake Oil and Gas Company assisted with collections. Finally, we thank all those who volunteered to assist with collections of Snow Geese (see list in Jónsson 2005). Any use of trade, firm, or product names is for descriptive purposes only and does not imply endorsement by the U.S. Government.

References

Afton, A.D. & Ankney, C.D. 1991. Nutrient-reserve dynamics of breeding Lesser Scaup: a test of competing hypotheses. *Condor* 93: 89–97.

Alisauskas, R.T. 1988. Nutrient reserves of Lesser Snow Geese during winter and spring migration. Ph.D. thesis, University of Western Ontario, London, Ontario, Canada.

Alisauskas, R.T. 1998. Winter range expansion and relationships between landscape and morphometrics of midcontinent Lesser Snow Geese. *Auk* 115: 851–862.

Alisauskas, R.T., Ankney, C.D. & Klaas, E. E. 1988. Winter diets and nutrition of midcontinental Lesser Snow Geese. *Journal of Wildlife Management* 52: 403–414.

Anderson, D.R. 2008. *Model-based Inference in the Life Sciences: A Primer on Evidence.* Springer, New York, USA.

Ankney, C.D. & MacInnes, C.D. 1978. Nutrient reserves and reproductive performance of female Lesser Snow Geese. *Auk* 95: 459–471.

Ankney, C.D. & Afton, A.D. 1988. Bioenergetics of breeding Northern Shovelers: diet, nutrient reserves, clutch size and incubation. *Condor* 90: 459–472.

Bishop, C.M., Butler, P.J.. El Haj, A.J. & Egginton, S. 1998. Comparative development in captive and migratory populations of the Barnacle Goose. *Physiological Zoology* 71: 198–207.

Bolen, E.G. & Rylander, M.K. 1978. Feeding adaptations in the Lesser Snow Goose (*Anser caerulescens*). *Southwestern Naturalist* 23: 158–161.

Buchsbaum, R., Wilson, J. & Valiela, I. 1986. Digestibility of plant constituents by Canada Geese and Atlantic Brant. *Ecology* 67: 386–393.

Burnham, K.P. & Anderson, D.R. 2002. Model selection and multi-model inference. Second edition. Springer, New York, USA.

Cooch, E.G., Lank, D.B., Dzubin, A., Rockwell, R.F. & Cooke, F. 1991. Body size variation in Lesser Snow Geese: Environmental plasticity in gosling growth rates. *Ecology* 72: 503–512.

Davies, J.C., Rockwell, R.F., & Cooke, F. 1988. Body size variation and fitness components in Lesser Snow Geese (*Chen caerulescens caerulescens*). *Auk* 105: 639–648.

Gauthier, G. & Tardif, J. 1991. Female feeding and male vigilance during nesting in Greater Snow Geese. *Condor* 93: 701–711.

Flint, P.L., Fowler, A.C., Bottitta, G.E. & Schamber, J. 1998. Observations of geese foraging for clam shells during spring on the Yukon-Kuskokwim Delta, Alaska. *Wilson Bulletin* 110: 411–413.

Fowler, A.C. & Ely, C.R. 1997. Behavior of Cackling Canada Geese during brood rearing. *Condor* 99: 406–412.

Halse, S.A. 1984. Diet, body condition, and gut size of Egyptian Geese. *Journal of Wildlife Management* 48: 569–573.

Harris, G.A. 1990. Grit site use by wildlife in southwestern Louisiana and southeastern Texas. M.Sc. thesis, Louisiana State University, Baton Rouge, Louisiana, USA.

Fox, A.D. & Kahlert, J. 2005. Changes in body mass and organ size during wing moult in non-breeding greylag geese *Anser anser*. *Journal of Avian Biology* 36: 538–548.

Gauthier, G., Bédard, J. & Bédard, Y. 1984. Comparison of daily energy expenditure of Greater Snow Geese between two habitats. *Canadian Journal of Zoology* 62: 1304–1307.

Jones, S.K.C., Cowieson, A.J., Williamson, S.A. & Munn, A.J. 2013. No effect of short-term exposure to high-fibre diets on the gastrointestinal morphology of Layer Hens (*Gallus gallus domesticus*): body reserves are used to manage energy deficits in favour of phenotypic plasticity. *Journal of Animal Physiology and Animal Nutrition* 97: 868–877.

Jónsson, J.E. 2005. Effects of body size and habitat use on goose behavior: Lesser Snow Goose and Ross's Goose. Ph.D. thesis, School of Renewable Natural Resources, Louisiana State University, Baton Rouge, Louisiana, USA. http://digitalcommons.lsu.edu/gradschool_dissertations/409.

Jónsson, J.E. & Afton, A.D. 2006. Differing time and energy budgets of Lesser Snow Geese in rice-prairies and coastal marshes in southwest Louisiana. *Waterbirds* 29: 451–458.

Jónsson, J.E. & Afton, A.D. 2015a. Does the proportion of Snow Geese using coastal marshes in southwest Louisiana vary in relation to light goose harvest or rice production? *Goose Bulletin* 20: 7–19.

Jónsson, J.E. & Afton, A.D. 2015b. Are Wintering Areas Shifting North? Learning from Lesser Snow Geese Banded in Southwest Louisiana. *Southeastern Naturalist* 14: 219–307.

Jónsson, J.E. & Afton, A.D. 2016. Do foraging methods in winter affect morphology during growth in juvenile Snow Geese? *Ecology and Evolution* 6: 7656–7670.

Jónsson, J.E., Frederiksen, M. & Afton, A.D. 2014. Movements and survival of Lesser Snow Geese *Chen caerulescens caerulescens* wintering in two habitats along the Gulf Coast, Louisiana. *Wildfowl* 64: 54–74.

Larsson, K., & Forslund, P. 1991. Environmentally induced morphological variation in the Barnacle Goose, *Branta leucopsis*. *Journal of Evolutionary Biology* 4: 619–636.

Laursen, K. & Møller, A.P. 2016. Your tools disappear when you stop eating: phenotypic variation in gizzard mass of eiders. *Journal of Zoology* 229: 213–220.

Mattocks, J.G. 1971. Goose feeding and cellulose digestion. *Wildfowl* 22: 107–113.

Mazerolle, M.J. 2015. AICcmodavg: model selection and multimodel inference based on (Q)AIC(c). R package version 2.0–3. http://CRAN.R-project.org/package=AICcmodavg.

Miller, M.R. 1975. Gut morphology of Mallards in relation to diet quality. *Journal of Wildlife Management* 39: 168–173.

Meixell, B.W., Arnold, T.W., Lindberg, M.S., Smith, M.R., Runstadler, J.A. & Ramey, A.M. 2016. Detection, prevalence, and transmission of avian hematozoa in waterfowl at the Arctic/sub-Arctic interface: co-infections, viral interactions, and sources of variation. *Parasites & Vectors* 9: 390.

Moorman, T.E., Baldassarre, G.A. & Richard, D.M. 1992. Carcass mass, composition and gut morphology dynamics of Mottled Ducks in fall and winter in Louisiana. *Condor* 94: 407–417.

Owen, M. 1980. *Wild Geese of the World*. B.T. Batsford Ltd., London, UK.

Paulus, S.L. 1982. Gut morphology of Gadwalls in Louisiana in winter. *Journal of Wildlife Management* 46: 483–489.

Piersma, T. & Lindström, A. 1997. Rapid reversible changes in organ size as a component of adaptive behaviour. *Trends in Ecology and Evolution* 12: 134–138.

Piersma, T., Gudmundsson, G.A. & Lilliendahl, K. 1999. Rapid changes in size of different functional organ and muscle groups in a long-distance migrating shorebird. *Physiological and Biochemical Zoology* 72: 405–415.

Prop, J. & Vulink, T. 1992. Digestion by Barnacle Geese in the annual cycle: the interplay between retention time and food quality. *Functional Ecology* 6: 180–189.

Relyea, R.A. 2004. Fine-tuned phenotypes: Tadpole plasticity under 16 combinations of predators and competitors. *Ecology* 85: 172–179.

Sedinger, J.S. 1997. Adaptations to and consequences of an herbivorous diet in grouse and waterfowl. *Condor* 99: 314–326.

Sedinger, J.S. & Raveling, D.G. 1988. Foraging behavior of Cackling Canada Goose goslings: implications for the roles of food availability and processing rate. *Oecologia* 75: 119–124.

Starck, J.M. 1999. Phenotypic flexibility of the avian gizzard: rapid, reversible and repeated changes of organ size in response to changes in dietary fibre content. *Journal of Experimental Biology* 202: 3171–3179.

Tieleman, B.I., Williams, J.B., Buschur, M.E. & Brown, C.R. 2003. Phenotypic variation of larks along an aridity gradient: are desert birds more flexible? *Ecology* 84: 1800–1815.

Thompson, J.E. & Drobney, R.D. 1996. Nutritional implications of molt in male Canvasbacks: variation in nutrient reserves and digestive tract morphology. *Condor* 98: 512–526.

van Gils, J.A., Piersma, T., Dekinga, A. & Dietz, M.W. 2003. Cost-benefit analysis of mollusc eating in a shorebird. II. Optimizing gizzard size in the face of seasonal demands. *Journal of Experimental Biology* 206(Pt 19): 3369–3380.

Williamson, S.A., Jones, S.K.C. & Munn, A.J. 2014. Is gastrointestinal plasticity in King Quail (*Coturnix chinensis*) elicited by diet-fibre or diet-energy dilution? *Journal of Experimental Biology* 217: 1839–1842.

Appendix 1. Overall body size (indexed by PC1 from a principal components analysis), for adult (n = 17 males, 23 females) and juvenile (n = 34 males, 36 females) Snow Goose specimens collected in southwest Louisiana, USA, during winters 2001–2004. Least Squares Means test for an age * sex interaction effect on the PC1 values found that adult males were significantly larger in size (lsmean ± s.e. = 3.12 ± 0.44, P < 0.0001 for all group comparisons) in comparison with the other groups (lsmean = –1.09 ± 0.37 for adult females, –1.23 ± 0.30 for juvenile females, and 0.48 ± 0.31 for juvenile males). Juvenile males were intermediate between adult males and the two female groups (P < 0.002 on comparison with adult females, P < 0.0001 on comparison with adult males, and P = 0.0001 on comparison with juvenile females). The two female groups did not differ from each other (P = 0.77, n.s.).

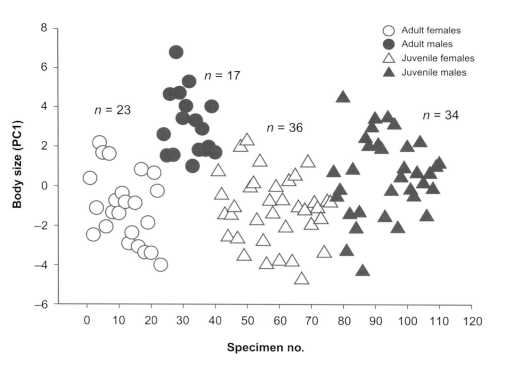

Allocation of parental care by Western Canada Geese *Branta canadensis moffitti*

KENNETH M. GRIGGS[1,2] & JEFFREY M. BLACK[1]*

[1]Waterfowl Ecology Research Group, Department of Wildlife, Humboldt State University, Arcata, California, USA.
[2]Klamath National Wildlife Refuge Complex, 4009 Hill Rd, Tulelake, California 96134, USA.
* Correspondence author. E-mail: Jeff.Black@humboldt.edu

Abstract

The level of parental care provided by Western Canada Geese *Branta canadensis moffitti* to their goslings (4–11 months of age) was measured in a resident population by determining proximity of goslings to the nearest parent, goslings' daily attendance in the family unit, and the duration of the parent-gosling association during the first winter. Time spent in vigilance postures (watching for competitors and predators) and aggression (to maintain space for foraging within flocks) was determined for each family member. Male goslings were more "helpful" in that they were more vigilant and aggressive than female goslings. Perhaps as a result, male goslings benefited more from all three measures of parental care than female goslings. Male goslings were on average closer to parents, in attendance more often, and had a longer duration of parent-gosling association during the first winter than their female siblings. Among females in the same brood (*i.e.* siblings), the most vigilant and aggressive were allocated more care as measured by proximity to parents, daily attendance, and duration of association with parents. Among male siblings, the most vigilant individuals were allocated with more care in terms of proximity to parents than less vigilant male goslings. Within sexes, gosling structural size (*i.e.* skull length) did not affect the allocation of parental care. With regard to parents, the level of female vigilance and aggression towards flock members was negatively correlated with the amount of "help" provided by the most "helpful" gosling in the brood, in terms of the goslings' contribution to the family through their vigilance and aggressive behaviours. This finding suggests that female parents benefit from maintaining contact with "helpful" goslings, more so than females with less "helpful" goslings. This relationship was not apparent for male parents. The most interesting finding from this study was that parent geese appeared to base parental investment decisions more on their goslings' behaviour rather than structural size. The energetic costs that mature goslings bear from assisting parents with family duties of watching for competitors and predators and defending foraging space within flocks may be compensated by longer-term benefits of prolonged association with their parents.

Key words: aggression, brood, Canada Geese, family, goslings, offspring, parental care, social, vigilance.

The mating system of wild geese and swans is characterised by: (1) long-term pair bonds, (2) biparental care, and (3) extended parent-offspring association (Kear 1970). In winter, goose and swan flocks typically consist of a combination of families, paired birds, and single subadults or adults prospecting for a mate (Boyd 1953; Evans 1979; Owen 1980; Scott 1980a). Family units include the parental pair and their offspring from the previous summer (Boyd 1953; Raveling 1970). Parents spend much of their day being vigilant to the threat of competitors or predators, and on defending space within the flock to enable disturbance-free foraging for their offspring (*i.e.* parental care: Scott 1980b,c; Black & Owen 1989a,b). The extra time that parents spend on vigilance and aggressive behaviour takes away from other essential daily activities, and for several species in the Anserinea subfamily has been used as a quantifiable measurement of parental investment in their young (Lazarus & Inglis 1978; Scott 1980b; Akesson & Raveling 1982; Black & Owen 1989a).

The timing of family break up in geese varies within and among species (Owen 1980), ranging from less than a full year for Cackling Geese *Branta hutchinsii minima* (Johnson & Raveling 1988) to remaining with parents over several winters for Greenland White-fronted Geese *Anser albifrons flavirostris* (Warren *et al.* 1993). In Barnacle Geese *Branta leucopsis*, most parents are seen with goslings in the first month after return to the wintering grounds (*i.e.* when the goslings were 4 months old), to a lesser degree during early winter (age 5–9 months), and even fewer parents continue to associate with their goslings in spring and during the return migration (10–11 months; Black & Owen 1989a). Barnacle Goose parents threaten and peck their offspring with increasing regularity during mid–late winter, suggesting that some (but not all) goslings leave the family because of this harassment (Black & Owen 1989a). As families become smaller, parental effort in vigilance and defence of foraging space eventually declines as mature goslings increase their participation in these behaviours (Black & Owen 1989a,b). This observation led to the suggestion that gosling "help" may enable parents to acquire essential body stores for the coming breeding season (Black & Owen 1989a). Furthermore, parents that maintained their association with at least one gosling for the longest period into the spring, bred more successfully in the following summer compared to those with shorter associations with goslings (Black & Owen 1989a). These observations lead to the question of how parents decide which goslings to expel and which to keep in the family. For example, do parents favour more "helpful" offspring?

Considering the parent-offspring relationship from the perspective of offspring in goose flocks, goslings may benefit from continued association with parents by increased social status and gaining access to prime feeding sites compared to goslings that curtail association with parents (Raveling 1970; Black & Owen 1984, 1987, 1989a,b; Sirwardina & Black 1999; Raveling *et al.* 2000). These benefits are similar to those described for Bewick's Swan *Cygnus columbianus bewickii* families, where cygnets in closest proximity to parents fed more, were threatened less by neighbours, and were more successful in aggressive encounters than individuals farther from parents (Scott 1980b). In Barnacle Geese, males that bred at least once during their lifetime spent significantly longer periods with their parents during their first winter than males that did not breed; there was no such difference for females (Black *et al.* 2014). These observations led us to question whether different goslings are allocated more than their share of the benefits of parental care (*i.e.* sibling rivalry, *sensu* Mock & Parker 1997). Theoretical arguments suggest that under variable environmental and ecological conditions, parents may extend unequal care to individual offspring (Winkler 1987; Clutton-Brock 1991), and that parents are expected to invest more in offspring with characteristics that result in higher inclusive fitness (Trivers 1972). For example, in some systems it might pay parents to invest more in the smallest, or weakest offspring in greatest need (Lessells 2002).

Assuming that young geese and swans gain substantial benefits from their association with parents, researchers have measured the allocation of parental care in terms of nearness or proximity to parents,

daily attendance in the family, and the duration of the parent-offspring association during the first year (Scott 1980b; Black & Owen 1989a; Black *et al.* 2014). The aim of this paper is to contribute to understanding how parental care is allocated within goose families. Using Western Canada Goose *Branta canadensis moffitti* broods as our model system, we examined whether parents provided more care to more "helpful" offspring or whether the amount of care was distributed in relation to goslings' relative "need". To this end, we tested whether the characteristics of individual offspring (*e.g.* sex, behaviour, and structural size) influenced allocation of parental care (*sensu* Clutton-Brock *et al.* 1981; Stamps *et al.* 1985). We also determined whether parents experienced any noticeable benefit from maintaining contact with "helpful" goslings by quantifying the change in parental effort devoted to maintaining space around family members (*sensu* Black & Owen 1989a).

Methods

This study was conducted on the pastures and saltmarshes adjacent to Humboldt Bay (40°47'44"N, 124°7'7"W) in northwest California, USA. The Arcata Bottomlands to the north of Humboldt Bay is comprised primarily of pastures managed for dairy cattle, sheep and cattle. The Humboldt Bay National Wildlife Refuge (HBNWR) to the south contains permanent and seasonal wetlands, saltmarsh and hay fields.

The study population of about 1,500 Western Canada Geese used these habitats during winter, breeding and brood-rearing (Griggs & Black 2004). In June 2000, 97 adult geese and 192 goslings were captured while flightless using a corral trap (*sensu* Cooch 1953). Birds were aged, sexed through cloacal examination, fitted with a U.S. Fish and Wildlife Service metal leg-ring and an alpha-coded plastic neck-collar, then weighed and measured (skull length; Dzubin & Cooch 1992). Most goslings hatched in April 2000, although exact hatch dates were not determined.

From 1 July 2000, after most goslings had fledged and parents completed their wing moult, through to the end of March 2001, flocks with collared individuals were observed with a spotting scope (Leica® 20–60×) to determine the identity of individuals within families. Average monthly flock sizes during winter ranged from 30–156 birds, peaking in November (Griggs & Black 2004). Family membership was based on the proximity of individuals, similar travel paths, mutual social displays, coordination of vigilance routines and assistance in aggressive encounters (Akesson & Raveling 1982; Black *et al.* 1996). Observations were conducted 3–5 times per week between 08:00–12:00 h (96% of dataset) and 13:00–15:00 h (4%) using a vehicle as a blind along farm roads within the study area. The average size of Western Canada Goose families on Humboldt Bay was 4.2 goslings (*s.e.* ± 0.7, *n* =35), including one family with a single gosling and one large family with 22 goslings (*i.e.* brood amalgamation). The following assessment was based on 24 families that were regularly observed (see below) with 2–10 goslings assumed to be "natural" family members (*i.e.* not adopted). Parents and goslings eventually separated when parents established nesting territories and goslings joined non-breeding flocks. Goslings were considered to be no longer associating with their parents when they were observed in a flock which did not include their parents on at least two occasions. Dates were assigned to three seasons: autumn (1 August–22 October 2000), early winter (23 October–31 December 2000), and late winter (1 January–15 March 2001).

Focal animal sampling was used to record the behavioural activity budget of parents and goslings (Martin & Bateson 1993), but our analysis in this paper focused on vigilance and aggression. At 30 s intervals, during 10 min sampling periods, the behaviour of each family member was recorded as vigilant (head and neck at an angle of > 45°) or as other daily activities. These data were converted to the proportion of intervals in vigilant posture for each bird, which was then averaged for each seasonal period (autumn, early winter and late winter). Behaviour sampling (continuous recording of conspicuous behaviours; Martin & Bateson 1993) was used to record the occurrence of initiating aggressive encounters for each family member during the entire period of observation; the initiator of aggressive encounters usually won the encounter (Boyd 1953; Black & Owen 1987, 1989b).

As goslings matured, three measures were used to describe the level of allocated parental care: (1) proximity of each gosling to nearest parent, (2) attendance (presence/absence) of each gosling in the family group, and (3) length of association with parents. These measures were based on the assumption that benefits of protection and learning opportunities were acquired when goslings were associating with their parents (*sensu* Lazarus & Inglis 1978; Raveling 1970, 1981; Scott 1980b; Black & Owen 1984, 1989a, b). The proximity of each gosling to each parent was recorded in goose-lengths (~90 cm, Bellrose 1980) at the beginning and end of each 10 min observation and averaged. This assessment was practised using goose decoys separated by different distances and angles prior to fieldwork in order to ensure accurate measures. Attendance in the family group was recorded as the proportion of days in which each gosling was seen to be associating with parents in the same flock up to the time that it was found to have left the family. The duration of parent-gosling association during the first year was calculated as the number of days from 1 August 2000 to the mid-point between date on which parents were last observed with a particular gosling and the first record without that gosling (Black & Owen 1989a).

In all analyses, indices of parental care were assessed in terms of proximity to parents, gosling attendance in the family, and the duration of the parent-offspring bond prior to family break-up. Indices of gosling "help" included their contribution to maintaining family status and position in flocks (*i.e.* goslings' vigilance and aggressiveness). For an index of goslings' relative "need" (see page 46), we used a measure of structural size (*i.e.* skull length). Sexes were analysed separately because male geese were generally larger, more vigilant and aggressive than females (Akesson & Raveling 1982; Black & Owen 1987, 1989a,b). All comparisons were conducted among siblings within broods. This removed

the need to control for gosling age among broods and reduced bias due to parent quality. After testing and confirming normality and equal variance in the three measures of parental care (proximity to parents, gosling attendance, family duration) we use ANOVA to test for variation in these measures over the three times of year (autumn, early winter and late winter).

To determine whether the amount of parental care (*i.e.* proximity to parents, attendance in the family, and duration with parents) was attributed unequally among goslings, we created two gosling categories from among siblings of the same sex. Category I (highest care) was assigned to the individual of each sex that received the very highest level of care in each brood. Category II (low care) was assigned to siblings of the same sex that received substantially less parental care – *i.e.* more than half a standard deviation less than the average value for all siblings of that sex in the brood. Families were not included in this analysis if these criteria were not met, thus reducing the sample of 24 families to *e.g.* 10, 7 and 5 families for the analysis of proximity to parents, attendance in the family and duration with parents, respectively, for comparison among female siblings. We used mean values for goslings' attributes (behaviour and size, see below) when multiple siblings of the same sex were included in Category II. Non-parametric Wilcoxon signed-rank tests were used to test for differences between Category I (highest care) and Category II (low care) in terms of the goslings' "helpful" behaviours (*i.e.* proportion vigilant and rate of aggression) and structural size (*i.e.* skull length).

Kruskal-Wallis tests were used to assess variation across the three season categories (autumn, early winter, late winter) in proportion of time spent performing different behaviours (*i.e.* proportion vigilant and rate of aggression).

To determine whether parents appeared to reduce their effort in terms of vigilance and aggression in relation to gosling "help" with these behaviours, we used Spearman Rank Correlation tests of parents' and goslings' behaviours. For this analysis, we calculated the change in parents' vigilance and aggression from early winter to late winter. For goslings, we calculated mean vigilance and rate of aggression from observations taken during early and late winter periods, and used values from the single most "helpful" gosling in each brood (*i.e.* the highest level of vigilance and aggression). Using these criteria reduced the sample of 24 families to 14 and 17, respectively, for the tests of vigilance and aggression.

Results

Western Canada Goose families were resighted and behaviours recorded 9–24 times during the course of the study (mean ± s.e. = 18.8 ± 0.9 resightings per family). To assess whether the parent-offspring associations changed over time, we compared goslings' average proximity to their parents and also family attendance in autumn, early winter and late winter. Gosling proximity to parents was greatest (ANOVA, males: $F_{2,123}$ = 3.49, P = 0.03; females: $F_{2,115}$ = 11.78, P < 0.001), and family attendance was lowest (males: $F_{2,124}$ = 32.46, P < 0.001; females: $F_{2,116}$ = 62.0, P < 0.001) in the late winter period

(Fig. 1). Parent geese established nest sites and territories on levies and islands in the Humboldt Bay area soon after this late winter period (Griggs & Black 2004), which is when the 10–11 month old goslings were observed in flocks in adjacent pastures without their parents.

To assess whether goslings increased their contribution towards maintaining foraging space within flocks, we compared goslings' behaviour across different times of year when goslings and parents were still together in families. The proportion of time that goslings spent being vigilant was greatest in late winter (Kruskal-Wallis, males: H_2 = 10.73, P = 0.005; females: H_2 = 9.30, P = 0.01) (Fig. 2). Levels of aggression (measured as frequency/h) were greatest among male goslings in early winter (H_2 = 15.9, P < 0.001); although female goslings followed a similar trend, the difference between the three time periods was not significant (H_2 = 2.15, n.s.; Fig. 2).

There was no significant difference in vigilance behaviour among different times of year in male or female parents. Aggression levels were greatest in early winter in adult females (H_2 = 7.78, P = 0.02), but did not differ significantly across time periods for adult males.

On comparing the association between male and female siblings with their parents, male goslings were found to be in closer proximity to their parents, had higher levels of attendance in the family, and had a longer length of association with parents than their female siblings (Table 1). This suggests that males received more of the benefits than their female siblings from the parent-offspring association. Males were also

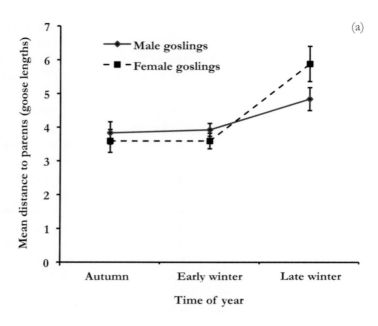

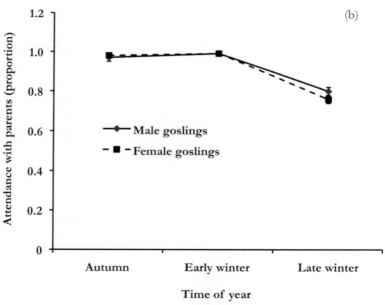

Figure 1. (a) Average distance from parents in goose lengths, and (b) attendance with parents (proportion of resightings where associating with parents), recorded for male and female Western Canada goose goslings during three time periods. Error bars are +/− one standard error.

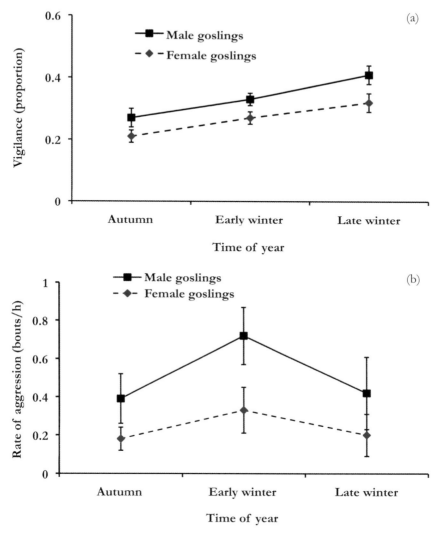

Figure 2. (a) Average proportion of time vigilant and (b) rates of aggression toward flock members in male and female Western Canada Goose goslings during 3 periods throughout the non-breeding season. Error bars are +/− one standard error.

significantly larger, more vigilant and more aggressive than female siblings (Table 2).

Within-sex comparisons among siblings indicated that parents may have allocated care based on behavioural rather than structural size. For example, Category I females (classed as having the highest level of parental care) were significantly more vigilant than Category II (low care) females (Wilcoxon signed-rank tests for each of the

Table 1. Levels of parental care provided to male and female goslings within broods of Western Canada Geese at Humboldt Bay, California from August 2000–March 2001. [a]Z and P values are the results of Wilcoxon signed-rank tests for differences in parental care accorded to male and female goslings. [b]Proximity = goslings' average distance to parents (in goose-lengths). [c]Attendance = gosling attendance with parents (proportion of resightings). [d]Duration = gosling duration with parents (in days from 1 August).

	Male			Female			Z[a]	P
	Mean	s.e.	n	Mean	s.e	n		
Proximity[b]	4.36	0.43	18	5.38	0.62	18	2.20	0.027
Attendance[c]	0.80	0.49	11	0.69	0.05	11	2.31	0.021
Duration (days)[d]	216.3	5.23	7	200.3	5.08	7	2.20	0.028

Table 2. Differences in the behavioural and physical characteristics of male and female goslings within broods of Western Canada Geese in the Humboldt Bay Area, August 2000–March 2001. Measures of aggression were toward individuals outside of family. Z and P values are the results of Wilcoxon signed-rank tests for differences between the sexes for each measure.

	Male			Female			Z	P
	Mean	s.e.	n	Mean	s.e.	n		
Proportion time vigilant	0.33	0.02	18	0.29	0.02	18	2.11	0.035
Rate of aggression (bouts/h)	0.64	0.14	18	0.4	0.09	18	2.35	0.019
Skull length (mm)	109.8	1.16	18	107.5	1.38	18	2.24	0.025

three measures of parental care: proximity $Z = 2.80$, $n = 10$, $P = 0.005$; attendance $Z = 2.37$, $n = 7$, $P < 0.02$; duration $Z = 2.02$, $n = 5$, $P < 0.05$; Fig. 3). Category I (highest care) females were also more aggressive than Category II (low care) females (proximity $Z = 2.39$, $n = 10$, $P < 0.02$; attendance $Z = 1.96$, $n = 7$, $P = 0.05$; duration $Z = 2.03$, $n = 5$, $P < 0.05$) (Fig. 3). This trend was similar on comparing vigilance levels for Category I and II males, but only one of the three measures of parental care was statistically significant (proximity to parents: $Z = 2.34$, $n = 20$, $P < 0.02$). Category I

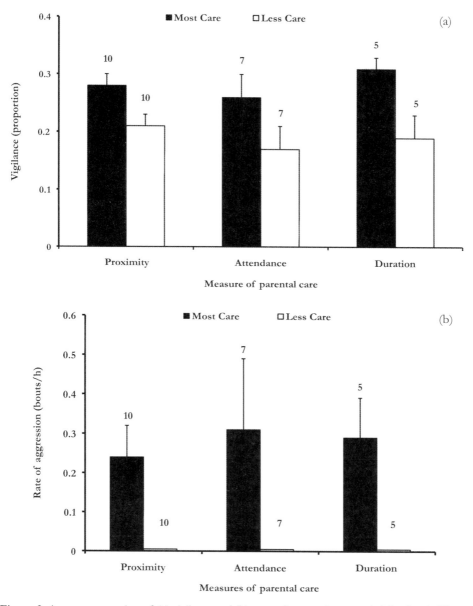

Figure 3. Average proportion of (a) vigilance and (b) rates of aggression recorded for female Western Canada Goose goslings receiving the highest amount and low amounts of parental care. Parental care was based on three measures: proximity to parents, attendance rate in family, and duration with parents into the spring. Error bars are one standard error. Sample sizes (number of families) are shown above the s.e. bars.

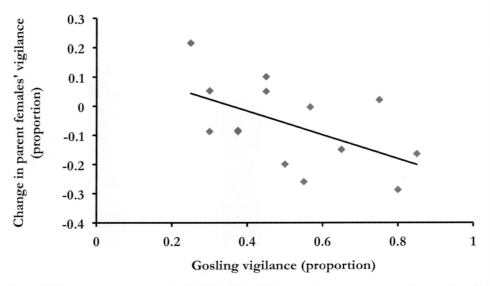

Figure 4. Relationship between the change in the proportion of time vigilant for Western Canada Goose parent females between early and late winter, and their goslings' vigilance in late winter. The mean proportion of time vigilant from the most vigilant gosling in each brood was used in the analysis.

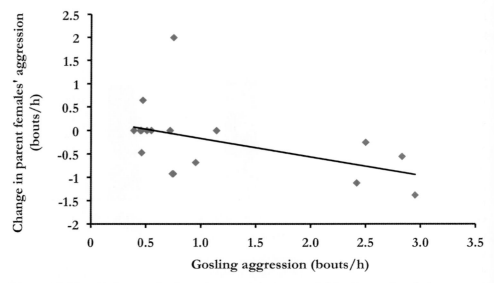

Figure 5. Relationship between the change in aggression rates recorded for Western Canada Goose parent females between early and late winter, and their goslings' aggression rates during the late winter period. The mean rate of aggression from the most aggressive gosling in each brood was used in the analysis.

(highest care) males were not significantly more aggressive than Category II (low care) males (proximity $Z = 1.14$, $n = 10$, n.s.; attendance $Z = 0.09$, $n = 7$, n.s.; no test for duration variable, $n = 1$). With regard to siblings' structural size, there was no significant difference in skull length between Category I and II goslings for either sex (range of results for females, for proximity, attendance and duration: $Z = 0.13–0.68$, $n = 5–10$, n.s.; range of results for males, for proximity and attendance: $Z = 0.66–0.68$, $n = 8–10$, n.s).

To test whether parents gained any noticeable benefit from maintaining contact with "helpful" goslings, we quantified the change in the parents' vigilance and aggression in relation to gosling "help" with these behaviours for each family. The change in the female parents' level of vigilance and aggression in late winter was negatively correlated with most "helpful" goslings' level of vigilance and aggression (vigilance, $r_s = –0.58$, $n = 14$, $P = 0.03$, Fig. 4; aggression, $r_s = –0.56$, $n = 17$, $P = 0.01$, Fig. 5). These relationships were not found for male parents and their goslings.

Discussion

This study contributes to our understanding of how parental care is allocated within goose families. When both sexes were present in broods, parents maintained contact with male more than female offspring. This was notable for all three parental care measures: proximity, attendance and family duration through the first year, indicating that males received more of the benefits from parental care than their female siblings. In Barnacle Geese,

long-term benefits from longer periods of parental care experienced in the first year was observable in males, but not in females, on considering the individuals' survival and eventual reproductive success (Black *et al.* 2014). Raveling *et al.* (2000) similarly documented higher rates of survival for immature Giant Canada Geese *Branta canadensis maxima* which had continued to associate with family members, in comparison with single goslings that fended for themselves in winter flocks. When associating with parents, goslings may learn social and predator detection skills, diet preferences and intricate features of foraging areas, breeding colony attributes, and landscape features along migration routes (Owen 1980; Raveling 1981; Marshall & Black 1992; Black & Owen 1989a; Black *et al.* 2014). Goslings may also assume the dominance status of their parents through association with these reproductively successful adults (Black & Owen 1984, 1989a,b; Raveling 1970). Future studies could test in further detail the occurrence and mechanism of such social inheritance in goose flocks (*sensu* Raveling 1970; Black & Owen 1987).

We are not certain whether male goslings maintained proximity and stayed longer than females in the family group as a result of parental choice or because these male goslings pushed siblings away. However, since this (and other studies) showed that maturing male goslings were significantly larger, more vigilant and more aggressive than female siblings (Table 2), it is likely that a sibling-sibling hierarchy was established within families (*sensu* Black & Owen 1987). In Barnacle Geese, goslings experienced an increasing number of parental attacks, which came mostly from male parents, but also from female parents and dominant siblings (Black & Owen 1989a). The only way that goslings are able to withstand these attacks is to employ a submissive "greeting" posture that subdues the aggressor (Radesäter 1974). Goslings use this behaviour when they approach or get approached by a parent or dominant sibling. Black *et al.* (2014) provided limited, but compelling evidence that male Barnacle Goose goslings were consistently closer, and were attacked least by parents. Furthermore, Black & Owen (1987) described that a rank order existed among siblings within broods, where males were eventually dominant over females as males grew larger in structural size.

The question remains about whether parents favoured gosling males to females because males were more "helpful." To address this question, we compared the allocation of types of parental care among same-sex siblings. Within-sex comparisons among siblings suggest that parents may have allocated care based on behavioural rather than structural size. For female goslings, the more vigilant and aggressive individuals were provided more of all types of parental care (Fig. 3), and more vigilant males were provided more care in one measure (proximity to parents). Perhaps parents based investment decisions on a "threshold of helpfulness," where individuals displaying a certain level of vigilance and aggression were preferred. By nature, males tended to be more vigilant and aggressive than females, thus ensuring they were above this threshold, while only the

most vigilant and aggressive females reached that threshold. This aligns with the idea that the most helpful offspring were favoured by parents, where "help" was in the form of gosling contributions toward watching for competitors and predators and maintenance of foraging space within flocks.

In some earlier studies of birds and mammals, parents have been found to attribute more care to weaker or smaller individuals, whereas in others parents favoured the larger, stronger individuals (Stamps *et al.* 1985; Clutton Brock 1991; Slagvold 1997; Lessells 2002). Our own study found that measures of parental care in Western Canada Geese did not vary according to gosling structural size (*i.e* skull length) within each of the sexes.

Results from this study contribute to understanding why some parents maintain contact with goslings well into the late winter and spring, while attempting to rebuild body stores for the next breeding season (*sensu* Black & Owen 1989a). Female Western Canada Goose parents spent less of their day being vigilant and chasing conspecifics when their families contained helpful offspring. This was determined by calculating the change (or reduction) in female parent's vigilance and aggression as the maturing offspring increased their own effort in watching and chasing flock members. These finding are in line with predictions from the "*Gosling Helper Hypothesis*," which suggested that parents would benefit from gosling help (*sensu* Black & Owen 1989a). However, the largest benefit to parents measured to date was only for females which continued their association with the most vigilant and aggressive goslings in the family.

While some of the "resident" Western Canada Geese in this study have flown north to moult in summer (~20%; Griggs & Black 2004), most remained in the same general area around Humboldt Bay, California throughout the year. In all cases, the period of parent-offspring association at Humboldt Bay ended when parents began a new nesting attempt in early spring. This occurred when the mature goslings were 10–11 months old. In contrast, geese that migrate to northern breeding grounds may arrive there with mature goslings still in tow (*e.g.* Prevett & MacInnes 1980). In Greater White-fronted Geese *Anser albifrons*, mature offspring (yearlings) fitted with neck-collars have been resighted on their parents' territories (Ely 1979). Warren *et al.* (1993) suggested that yearlings may provide a form of "alloparental" care by helping to defend territorial boundaries on the breeding grounds. Fox *et al.* (1995) quantified time spent foraging and in vigilant postures for White-fronted Goose pairs on arrival prior to nest establishment, and found that pairs with yearlings still in association spent more time foraging and less time vigilant than pairs without yearlings, which lends further support to the "helper" hypothesis. The authors went on to propose (on page 155 of Fox *et al.* 1995) that maturing goslings experience a "developmental switch from offspring as dependents (eliciting additional vigilance in parents during their first summer) to offspring as cooperators (sharing vigilance with parents)."

Most research in other species have emphasised the costs of parental care (Trivers 1974; Parker 1985; Clutton Brock 1991). The results described in this paper provides evidence that parents may benefit from providing care to offspring with helpful characteristics. Perhaps the mutual benefits that both parents and goslings receive lessens the conflict that may arise with extended periods of parental care in goose flocks.

Acknowledgements

This study was funded by the California Department of Fish & Wildlife, California Duck Stamp Account, Humboldt State University Foundation, California Waterfowl Association, and the Nielsen Institute. For access to property, we thank California State Parks, the Cities of Arcata, Eureka and McKinleyville, the Humboldt Bay Harbor District, the Humboldt Bay National Wildlife Refuge, D. Hunt, R. Moxon, J. Nunes, D. Santos and the Green Diamond Resource Company. E. Bjerre, D. Goley, D. Kitchen, D. Lee, J. Moore and two anonymous referees reviewed the manuscript and provided statistical advice. K. Dawson, J. Long, K. Rogers and B. Yost were key members of the goose capture team.

References

Akesson, T.R. & Raveling, D.G. 1982. Behaviors associated with seasonal reproduction and long-term monogamy in Canada Geese. *Condor* 84: 188–196.

Bellrose, F.C. 1980. *Ducks, Geese and Swans of North America*. Stackpole Books, Harrisburg, Pennsylvania, USA.

Black, J.M. & Owen, M. 1984. The importance of the family unit to Barnacle Goose offspring: a progress report. *Norsk Polarinstitute Skrifter* 181: 79–85.

Black, J.M. & Owen, M. 1987. Determinant factors of social rank in goose flocks: acquisition of social rank in young geese. *Behaviour* 102: 129–146.

Black, J.M. & Owen, M. 1989a. Parent-offspring relationships in wintering Barnacle Geese. *Animal Behaviour* 37: 187–198.

Black, J.M. & Owen, M. 1989b. Agnostic behaviour in Barnacle Goose flocks: assessment, investment, and reproductive success. *Animal Behaviour* 37: 199–209.

Black, J.M., Choudhury, S. & Owen, M. 1996. Do Barnacle Geese benefit from lifelong monogamy? *In* J. M. Black (ed.), *Partnerships in Birds, the Study of Monogamy*, pp. 91–117. Oxford University Press, Oxford, UK.

Black, J., Prop, J. & Larsson, K. 2007. *Wild Goose Dilemmas: Population Consequences of Individual Decisions in Barnacle Geese*. Branta Press, Groningen, The Netherlands.

Black, J.M., Prop, J. & Larsson, K. 2014. *The Barnacle Goose*. T. & A. D. Poyser, London, UK.

Boyd, H. 1953. On encounters between wild White-fronted Geese in winter flocks. *Behaviour* 5: 85–129.

Clutton-Brock, T.H. 1991. *The Evolution of Parental Care*. Princeton University Press, Princeton, New Jersey, USA.

Clutton-Brock, T.H., Albon, S.D. & Guinness, F.E. 1981. Parental investment in male and female offspring in polygynous mammals. *Nature* 289: 487–489.

Cooch, G. 1953. Techniques for mass capture of flightless blue and Lesser Snow Geese. *Journal of Wildlife Management* 17: 460–465.

Dzubin, A. & Cooch, E.G. 1992. *Measurements of Geese: General Field Methods*. California Waterfowl Association, Sacramento, California, USA.

Ely, C.R. 1979. Breeding biology of the White-fronted Goose. M.Sc. thesis, University of California, Davis, USA.

Evans, M.E. 1979. Population composition, and return according to breeding status of Bewick's Swans wintering at Slimbridge, 1963 to 1976. *Wildfowl* 30: 118–128.

Fox, A.D., Boyd, H. & Bromley, R.G. 1995. Mutual benefits of associations between breeding and non-breeding White-fronted Geese *Anser albifrons*. *Ibis* 137: 151–156.

Griggs, K.M. & Black, J.M. 2004. Assessment of a Western Canada Goose translocation: landscape use, movement patterns and population viability. *In* T.J. Moser, R.D. Lein, K.C. VerCauteren, K.F. Abraham, D.E. Anderson, J.G. Bruggink, J.M. Coluccy, D.A. Graber, J.O. Leafloor, D.R. Luukkonen & R.E. Trost (eds.), *Proceedings of the 2003 International Canada Goose Symposium, 19–21 March 2003*, pp. 214–222. Madison, Wisconsin, USA.

Johnson, J.C. & Raveling, D.G. 1988. Weak family associations in Cackling Canada Geese during winter: effects of size and food resources on goose social organization. *In* M. Weller (ed.), *Waterfowl in Winter*, pp. 71–89. University of Minnesota Press, Minneapolis, Minnesota, USA.

Kear, J. 1970. Adaptive radiation of parental care in waterfowl. *In* J.H. Crook (ed.), *Social Behaviour in Birds and Mammals: Essays on the Social Ethology of Animals and Man*, pp. 357–92. Academic Press, New York, USA.

Lazarus, J. & Inglis, I.R. 1978. The breeding behaviour of the Pink-footed Goose: parental care and vigilant behaviour during the fledging period. *Behaviour* 65: 62–88.

Lessells, C.M. 2002. Partially biased favouritism: why should parents specialize in caring for different offspring? *Philosophical Transactions of the Royal Society, London, B.* 357: 381–403.

Marshall, A.M. & Black J.M. 1992. The effect of rearing experience on subsequent behaviour traits in captive Hawaiian Geese: implications for the re-introduction programme. *Bird Conservation International* 2: 131–147.

Martin, P.M. & Bateson, P. 1993. *Measuring Behaviour: An Introductory Guide.* Cambridge University Press, Cambridge, UK.

Mock, D.W. & Parker, G.A. 1997. *The Evolution of Sibling Rivalry.* Oxford University Press, Oxford, UK.

Owen, M. 1980. *Wild Geese of the World: Their Life History and Ecology.* B.T. Batsford, London, UK.

Parker, G.A. 1985. Models of parent-offspring conflict. V. Effects of the behaviour of the two parents. *Animal Behaviour* 33: 519–533.

Prevett, J.P. & MacInnes, C.D. 1980. Family and other social groups in Snow Geese. *Wildlife Monographs* 71: 1–46.

Radesäter, T. 1974. Form and sequential association between the Triumph Ceremony and other behaviour patterns in Canada Geese *Branta canadensis* L. *Ornis Scandinavica* 5: 87–101.

Raveling, D.G. 1970. Dominance relationships and agonistic behaviour of Canada Geese in winter. *Behaviour* 37: 291–319.

Raveling, D.G. 1981. Survival, experience, and age in relation to breeding success of Canada Geese. *Journal of Wildlife Management* 45: 817–829.

Raveling, D.G., Sedinger, J.S. & Johnson, D.S. 2000. Reproductive success and survival in relation to experience during the first two years in Canada Geese. *Condor* 102: 941–945.

Scott, D.K. 1980a. The behaviour of Bewick's Swans at the Welney Wildfowl Refuge, Norfolk, and on the surrounding fens: a comparison. *Wildfowl* 31: 5–18.

Scott, D.K. 1980b. Functional aspects of prolonged parental care in Bewick's Swans. *Animal Behaviour* 28: 938–952.

Scott, D.K. 1980c. Functional aspects of the pair bond in Bewick's Swans. *Behavioural Ecology and Sociobiology* 7: 323–327.

Slagvold, T. 1997 Brood division in birds in relation to offspring size: sibling rivalry and parental control. *Animal Behaviour* 54: 1357–1368.

Siriwardena, G.M. & Black, J.M. 1999. Parent and gosling strategies in wintering Barnacle Geese. *Wildfowl* 49: 18–26.

Stamps, J.A., Clark, A., Arrowood, P. & Kus, B. 1985. Parent-offspring conflict in budgerigars. *Behaviour* 94: 1–39.

Trivers, R.L. 1972. Parental investment and sexual selection. *In* B. Campbell (ed.), *Sexual Selection and the Descent of Man*, pp. 136–179. Aldine Publishing Company, Chicago, Illinois, USA.

Trivers, R.L. 1974. Parent-offspring conflict. *American Zoology* 14: 249–264.

Warren, S.M., Fox, A.D., Walsh, A. & O'Sullivan, P. 1993. Extended parent-offspring relationships in Greenland White-fronted Geese (*Anser albifrons flavirostris*). *Auk* 110: 145–148.

Winkler, D.W. 1987. A general model for parental care. *American Naturalist* 130: 526–543.

Photograph: Western Canada Goose female with brood, by Leslie Scopes Anderson.

How many Laysan Teal *Anas laysanensis* are on Midway Atoll? Methods for monitoring abundance after reintroduction

MICHELLE H. REYNOLDS[1], KAREN N. COURTOT[1] & JEFF S. HATFIELD[2]*

[1]US Geological Survey, Pacific Island Ecosystems Research Center, Kīlauea Field Station, Hawai'i National Park, Hawai'i 96718, USA.
[2]US Geological Survey, Patuxent Wildlife Research Center, Laurel, Maryland 20708, USA.
*Correspondence author. E-mail: jhatfield@usgs.gov

Abstract

Wildlife managers often request a simple approach to monitor the status of species of concern. In response to that need, we used eight years of monitoring data to estimate population size and test the validity of an index for monitoring accurately the abundance of reintroduced, endangered Laysan Teal *Anas laysanensis*. The population was established at Midway Atoll in the Hawaiian archipelago after 42 wild birds were translocated from Laysan Island during 2004–2005. We fitted 587 birds with unique markers during 2004–2015, recorded 21,309 sightings until March 2016, and conducted standardised survey counts during 2007–2015. A modified Lincoln-Petersen mark-resight estimator and ANCOVA models were used to test the relationship between survey counts, seasonal detectability, and population abundance. Differences were found between the breeding and non-breeding seasons in detection and how maximum counts recorded related to population estimates. The results showed strong, positive correlations between the seasonal maximum counts and population estimates. The ANCOVA models supported the use of standardised bi-monthly counts of unmarked birds as a valid index to monitor trends among years within a season at Midway Atoll. The translocated population increased to 661 adult and juvenile birds (95% CI = 608–714) by 2010, then declined by 38% between 2010 and 2012 after the Tōhoku Japan earthquake-generated tsunami inundated 41% of the atoll and triggered an Avian Botulism type C *Clostridium botulinum* outbreak. Following another severe botulism outbreak during 2015, the population experienced a 37% decline. Data indicated that the Midway Atoll population, like the founding Laysan Island population, is susceptible to catastrophic population declines. Consistent standardised monitoring using simple counts, in place of mark-recapture and resightings surveys, can be used to evaluate population status over the long-term. We estimate there were 314–435 Laysan Teal (95% CI for population estimate;

point estimate = 375 individuals) at Midway Atoll in 2015; *c.* 50% of the global population. In comparison, the most recent estimate for numbers on Laysan Island was of 339 individuals in 2012 (95% CI = 265–413). We suggest that this approach can be used to validate a survey index for any marked, reintroduced resident wildlife population.

Key words: Chapman estimate, count index, Laysan Teal, Lincoln-Petersen estimate, mark-resight models.

Species reintroduction programmes are being used increasingly to restore biodiversity and reduce extinction risk (Seddon *et al.* 2007; Miskelly & Powlesland 2013; Batson *et al.* 2015). Intensive post-release monitoring, such as radio-tracking founder birds, is important during the early post-release stage or initial breeding seasons to yield precise estimates of survival and reproduction. As a next step to understanding the outcome of a reintroduction attempt and to inform future management, marking a proportion of the population (*e.g.* with leg rings) facilitates monitoring individuals for estimating survival and abundance using capture-recapture or mark-resight analyses (Fischer & Lindenmayer 2000; Armstrong & Seddon 2008). After the population has increased and become established, a reduction in monitoring intensity might be warranted if systematic and accurate population data can be collected with reduced effort (Parker *et al.* 2013). Indices are often used to express comparisons of changes over a period of time and are often applied to infer population abundance from surveys of unmarked birds (*i.e.* direct counts). The valid application of a population index using survey counts requires testing of the assumption that the index is proportional to population size (Nichols 1992; White 2005).

In a previous study we tested the assumptions of survey monitoring protocols for providing a valid population index for Laysan Teal *Anas laysanensis* (classed as Critically Endangered globally; IUCN 2016) on Laysan Island. The Laysan Island study used a Lincoln-Petersen estimator to relate the survey counts to abundance derived from 15 years of mark-recapture and resightings data (Reynolds *et al.* 2015b). The reintroduced population at Midway Atoll National Wildlife Refuge was established with 42 birds translocated from Laysan Island during 2004–2005 (Reynolds *et al.* 2008), and the population increased rapidly to > 500 adult and juvenile birds by 2008 (Reynolds *et al.* 2011). In the study presented here, we estimate abundance over the period 2004–2015 for Laysan Teal reintroduced to Midway Atoll (hereafter, Midway), then provide linear regression equations for the 2007–2015 data to relate the maximum counts to estimates of abundance based on the Lincoln-Petersen estimator by season using analysis of covariance (ANCOVA). Our approach of transitioning from labour-intensive radio tracking, to less intensive mark-resight data, and then to the least intensive index of population abundance from unmarked birds, may be useful for other reintroductions, monitoring or restoration efforts. Our study using 8 years of data from marked Laysan Teal can serve as a model of how to validate a count index for other marked, resident wildlife populations undergoing systematic monitoring. Thus, this approach has utility for wildlife managers of reintroduced populations seeking to transition to a simple index to monitor population abundance for long-term trend analysis.

Methods

Study area

Midway is a remote Pacific atoll 2,300 km northwest of Honolulu (25°46'N, 171°44'W; Fig. 1) and is a part of the Papahānaumokuākea Marine National Monument (Executive Order 13022; Presidential Proclamation 8031 15 June 2006). The atoll consists of three islands

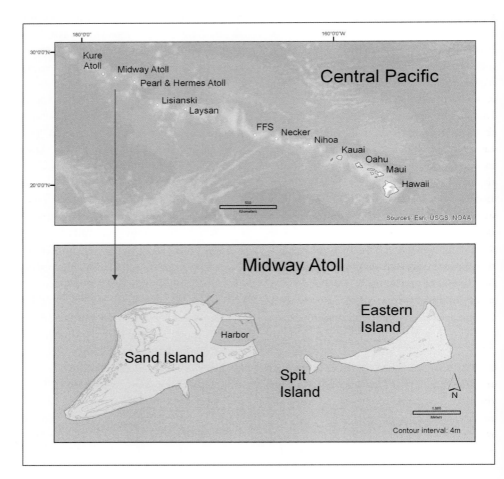

Figure 1. Map of the Hawaiian Islands, with detail of Midway Atoll.

otalling 604 ha, with a mean elevation of 3.0 m (Reynolds *et al.* 2015a).

Study species

The Laysan Teal is a formerly widespread Hawaiian dabbling duck that inhabited diverse habitats, but it had been confined to a small, remote low-lying atoll for about 150 years (Olson & Ziegler 1995). Bones of Laysan Teal are widespread and indicate that the species once occurred throughout the Hawaiian archipelago (Cooper *et al.* 1996) but, like many other Hawaiian birds (Olson & James 1991), it was extirpated from the larger islands following the arrival of humans and introduced rats 800–900 years ago (Burney *et al.* 2001). Unlike most of these species, the Laysan Teal survived on small islands in the remote northwestern chain, including Lisianski Island, from which it was extirpated in the mid-1800s, and Laysan Island, where the last relict population persisted. The Laysan Teal is non-migratory, primarily insectivorous and nests on the ground in dense terrestrial vegetation (see also Warner 1963; Reynolds *et al.* 2006, 2007, 2009). Wild-to-wild translocations have been made to Midway Atoll and to Kure Atoll, to reduce extinction risk and restore the species' range (USFWS 2009).

Population monitoring

Initially radio telemetry was used to monitor translocated individuals and their offspring during 2004–2007 (see Reynolds *et al.* 2008). As the population grew, however, resightings of individually marked birds, then standardised surveys were initiated. Laysan Teal have been caught on Midway during nine capture periods (May–December 2005, June 2006–February 2007, June–November 2007, August–November 2008, March–October 2009, October 2011, September–October 2013, August 2014 and July–December 2015; see also Reynolds *et al.* 2011), with most ducks captured using noose-carpet traps during crepuscular periods, or at night using a flexible hand-held net. Birds were marked with a numbered aluminium ring on one leg and a unique field-readable plastic engraved colour ring (Haggie Engraving, Crumpton, Maryland) or a field-readable engraved aluminium ring (Gey Band and Tag Company, Norristown, Pennsylvania) on the other leg.

The first standardised mark-resightings survey (hereafter survey) of Laysan Teal at Midway was conducted 23 October 2007, when collection of systematic survey data was discontinued until September 2008. Additional survey gaps occurred during February–May 2010 and December 2012–October 2014. During September 2008 to March 2016 surveys were typically conducted weekly or bi-monthly (see Reynolds *et al.* 2011 for detailed survey methods). Because the surveys and resightings were made weekly or bimonthly (*i.e.* over a relatively short time period), and most individually marked birds were sighted frequently on the small atoll (approximately once per week), the assumption of population closure is appropriate for mark-recapture models (see Reynolds *et al.* 2011 for detailed treatment of model assumptions). The survey started at sunrise and included all wetlands, persistent standing water, and freshwater guzzlers (*i.e.* water troughs) in the atoll. The start location and direction of survey routes were assigned randomly to reduce spatial-temporal bias. Observers recorded the ring status of each bird observed (as ringed, unringed or undetermined), and identified as many individual ringed birds as possible by colour-coded ring combinations or by reading the aluminium rings. Birds were classified as downy ducklings or post-fledglings (adult and feathered independent young of the year), with the population estimates including both juveniles and pre-breeders.

Data from uniquely marked individuals, identified from trapping, systematic surveys, incidental resightings and collection of carcasses, were used to determine the last date on which the birds were observed alive and to calculate each individual's lifetime median resighting frequency (*i.e.* 50th percentile of intervals between resightings). Each individual's median resightings interval was used to determine whether a missing bird (not seen again during the time series) was likely to be alive on a given survey date (details below).

Statistical analysis

Population estimates. Lincoln-Petersen (hereafter LP) based estimators have been used previously to estimate Laysan Teal abundance on Laysan Island and Midway (Moulton & Weller 1984; Marshall 1992; Reynolds & Citta 2007; Reynolds *et al.* 2011, 2015b). To estimate post-fledgling population abundance we used a mark-recapture sampling framework and a Chapman (1951) bias-corrected modification to the LP estimator:

$$\hat{N}_t = \frac{(M_t+1)(n_t+1)}{m_t+1} - 1$$

where $\hat{N}_t$ is the population estimate, M_t is the total marked population, n_t is the number of animals counted, and m_t is the number of marked animals counted (*i.e.* resighted), all at a given time t.

Data were divided into two periods to correspond with a breeding year: typical breeding (March–August, covering laying, incubation, brood rearing and moult) and typical non-breeding (September–February, with January and February grouped with the previous year, covering late moult, flocking, courtship, pairing and pre-breeding). Transition months may however need adjustment in future estimates to reflect actual breeding phenology for a given survey year, although such adjustment was not required in the current study. The post-fledgling population is geographically closed because there is no immigration or emigration between atolls, and demographically closed because timing of recapture-resighting periods was relatively short compared to the time interval between such periods. We removed individuals from total marked (M_t) live birds for the next survey if their marked carcass was recovered. Additionally, we estimated M_t using an individual's resightings history following Reynolds *et al.* (2011) and Reynolds *et al.* (2015b) to account for mortality of marked birds (M_t) where the carcass was not recovered. If a bird was not resighted after ringing it was excluded and assumed dead ($n = 12$). If a bird was not sighted after its median resightings interval, and never seen again, we assumed it was dead and excluded it from M_t at the next survey date. Thus, we inferred an individual's survival or mortality for each individually marked bird based on their individual resight frequency. This inferred mortality was calculated for every survey and every marked individual to better meet assumptions of the LP estimator.

Since LP estimators tend to overestimate population sizes, we used criteria, based on Robson & Regier (1964), to reduce overestimation bias and identify the highest quality survey for estimating abundance. These criteria were: highest counts within a period where the coefficient of variation (CV) of the LP estimator was < 10%, and where the percentage of teal known to be ringed or unringed (*i.e.* their ring status was known) identified during the survey was ≥ 60%. If multiple seasonal surveys met these criteria, we selected the count with the maximum percentage of known ducks.

Index validation. We used SAS v9.4 (SAS Institute Inc. 2012) to conduct analysis of covariance (ANCOVA) to investigate the relationship between the maximum of the observed counts (dependent variable) each season and year (hereafter maximum seasonal count) and the population abundance estimates for that survey date (independent variable). The ANCOVA allowed us to determine if the maximum count per season was a suitable index for population abundance and to test for differences between seasons. The full model had different slopes and intercepts for each season on comparing the linear relationship between the LP estimate and the maximum count. The reduced model (main effects: count and season, no interaction term) allowed for identical slopes for each season, but different intercepts (*i.e.* parallel

relationships). A further reduced model was examined that allowed for the same slope and intercepts across seasons (*i.e.* a simple linear regression, or correlation, of LP estimate *vs.* maximum count, pooling over the two periods). In addition, one survey was selected at random for each month in each year, and of those randomly selected counts, the maximum count per season was selected (hereafter maximum random count) and these data were analysed using ANCOVAs as described above. This model may be more applicable to future survey efforts by managers under funding limitations because it requires only one quality survey per month to generate maximum counts over seasons.

Results

Population counts and estimates

During 2004–2015, 587 Laysan Teal were fitted with unique leg rings and 21,309 recaptures and resightings were recorded through 07 March 2016. The median resightings interval across all individuals was eight days, and varied from one day for frequently seen birds to 399 days for a rarely seen bird. The median number of sightings per individual bird was 27 (range = 1–208). The maximum number of marked individuals alive in the population (339); 58% of the estimated 581 total birds (95% CI = 540–623), occurred in December 2009. When a ringed carcass was recovered (*n* = 139 reported), the median difference between the estimated date of death and actual carcass recovery was 47 days.

The highest rate of detection of post-fledglings occurred during the non-breeding period (see below); therefore we used counts from this period to estimate maximum annual post-fledgling abundance. We identified the best quality surveys, which were expected to yield the most accurate estimates to within 10–25% of the population abundance (Robson & Regier 1964); LP estimates in the non-breeding period ranged from 209 (95% CI = 185–232) in 2007, to 661 (95% CI = 608–714) in 2010 (Table 1, Fig. 2). A population decline of 38% occurred between the non-breeding seasons of 2010 and 2012 (Table 1). This

was observed after winter storms, followed by the Tōhoku tsunami in March 2011, and a Botulism type C *Clostridium botulinum* outbreak as a result of massive seabird die-offs from sudden flooding (Reynolds *et al.* 2017). By February 2015 adult and juvenile abundance grew to 599 (95% CI = 518–680); however, the population declined by 37% following another severe Botulism type C outbreak later that same year (Table 1, Fig. 2; USGS National Wildlife Health Center, Honolulu, Hawai'i, unpubl. data, 22 Mar–26 Sep 2011 and 21 Apr–27 Nov 2015).

Table 1. Maximum counts and modified Chapman bias-corrected Lincoln-Petersen mark-resight population abundance (95% confidence intervals; CI) for Laysan Teal at Midway Atoll, Hawai'i, for the years 2007–2015. The estimates shown are derived for the best survey during the non-breeding period that had the greatest proportion of teal with known ring status (*i.e.* know to be ringed or unringed) and met Chapman's (1951) standards for an unbiased estimator or criteria for marked samples sizes (Robson & Regier 1964). In all cases we chose best quality surveys, defined as being the highest count within a period where the coefficient of variation in relation to the LP estimator was < 10% and where the percentage of teal with known ring status identified during the survey was ≥ 60%.

Year	Count	Proportion of the population ringed	Proportion of ringed birds with known ring status	95% CI of population estimate
2007	135	0.43	0.90	185–232
2008	361	0.41	0.92	458–520
2009	349	0.52	0.95	508–571
2010	375	0.41	0.67	608–714
2011	263	0.59	0.92	369–414
2012	284	0.38	0.97	374–441
2013				No estimate
2014	352	0.19	0.95	518–680
2015	211	0.22	0.86	314–435

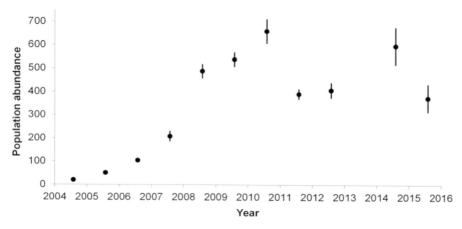

Figure 2. Modified Chapman bias-corrected Lincoln-Petersen mark-resightings population abundance estimates with 95% confidence intervals (CI) for Laysan Teal at Midway Atoll, Hawai'i, for the years 2007–2015. Before 2007 all adults were given radio transmitters, so the abundance was known exactly for those 3 years. No abundance estimate was generated for 2013 because no surveys were undertaken in this calendar year.

Index validation

The percentage of birds observed (*i.e.* the ratio of the maximum seasonal count to the LP estimate) averaged 62% (s.d. = 10%, range = 47–74%) during the 6-month non-breeding season and 33% (s.d. = 11%, range = 15–43%) during the 6-month breeding season. In the ANCOVAs the interaction of count and season was not significant ($P > 0.05$, n.s.), so the interaction term was dropped from the model. Models fitted with the main effects of year and season when comparing the direct counts to LP estimated counts explained most of the variability in the data (maximum seasonal count $r^2 = 0.77$, $n = 12$; maximum random count $r^2 = 0.81$, $n = 12$). The maximum seasonal count and the maximum random count models each had significant terms for count ($P = 0.0005$ and $P = 0.0004$, respectively) and season ($P = 0.0088$ and

$P = 0.0006$, respectively). Since statistical power and results of the ANCOVAs were qualitatively similar between analyses, we present results only from the maximum random count analysis. If we ignore season, the correlation between maximum counts and the LP estimates (or the slope in the simple linear regression) is not significant ($r = 0.05$, $n = 12$, $P = 0.09$, n.s.). The relatively large correlations shown in Fig. 3 imply that the counts within a season are a good index of population abundance.

A seasonal Laysan Teal population estimate (y), and the confidence bounds around the estimate ($t_{\alpha/2,n-2}\sqrt{MSE\ 1/n + \frac{(x-\bar{x})^2}{\Sigma_i(x_i-\bar{x})^2}}$; Sokal & Rohlf 1995), can be calculated based on a season's highest count (x) and season-specific equations from the most appropriate model (Fig. 3). Equations to estimate season-specific abundance apply only to seasonal maximum counts derived

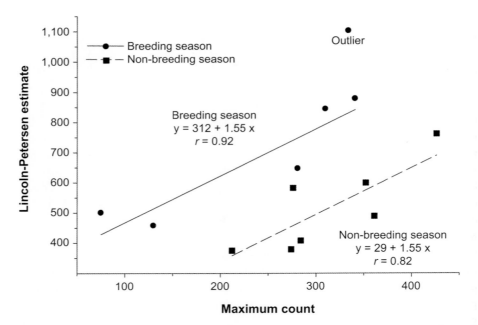

Figure 3. Modified Chapman bias-corrected Lincoln-Petersen estimates *vs.* maximum observed counts for Laysan Teal on Midway Atoll, Hawai'i, for the years 2007–2015 assuming equal slopes among seasons (*i.e.* reduced model). Seasons (breeding, non-breeding) are shown separately, with the regression line for each season also plotted separately. One outlier (LP = 1,100) was excluded from this analysis. Equations to estimate season-specific abundance apply only to seasonal maximum counts within the 8-year observed range of maximum counts (breeding = 75–341 birds; non-breeding = 212–426 birds).

from surveys conducted once or twice monthly within the observed count range (breeding season = 75–341 birds, non-breeding season = 212–426 birds); survey results outside of these ranges are not validated. Season-specific regression equations for deriving abundance from the maximum count and 95% confidence bounds are as follows:

Non-breeding:

$$y = 29.41 + 1.55x \pm$$
$$2.571\sqrt{6824.58 \ 1/7 + \frac{(x-312.14)^2}{30520.86}}$$

Breeding:

$$y = 312.09 + 1.55x \pm$$
$$3.182\sqrt{5633.08 \ 1/5 + \frac{(x-227.40)^2}{55313.20}}$$

Discussion

The progression for monitoring the translocated population first included intensive radio-tracking during the early post-release period (2004–2007), followed by population monitoring using mark-resightings and recapture data combined with systematic counts (in 2007–2015).

Now, a direct standardised survey count, without requiring the capture and marking of birds, will be a substantially less labour- and data- intensive approach for monitoring Laysan Teal abundance at Midway. A similar study undertaken on Laysan Island (Reynolds *et al.* 2015b) showed that the highest detection of birds also occurred in the non-breeding season (autumn and winter), and that both seasons showed a high correlation ($r = 0.82$–0.92) between estimated abundance and the maximum counts. The equations for the linear regressions, along with equations to estimate 95% confidence intervals, allow for a simpler survey approach that can utilise previous time series data from surveys and provide simpler analyses for managers than previously applied models (Marshall 1992; Reynolds & Citta 2007; Seavy *et al.* 2009).

The population at Midway Atoll grew to a total of 661 birds (95% CI = 608–714) in 2010, then a population decline of 38% was observed between 2010 and 2012 after the 2011 Tōhoku earthquake-generated tsunami. By 2014, the population had begun to recover from the tsunami (LP estimate = 599 birds, 95% CI = 518–680), but following a severe botulism outbreak during 2015 the population again experienced a 37% decline. Data indicate that the Midway population, like the founding Laysan Island population, is susceptible to catastrophic population declines (Seavy *et al.* 2009), and consistent standardised monitoring using simple counts can be used to evaluate population status over the long-term. For 2015, we estimated that there were 314–435 (95% CI for population estimate, point estimate = 375) teal on Midway

Atoll, or approximately 50% of the global population. In comparison, the 2012 estimate for Laysan Island was 339 individuals (95% CI = 265–413; Reynolds *et al.* 2015b).

Future surveys and estimates:

Our model relies on at least one or two high-quality atoll-wide surveys per month at Midway for this count index to have utility for estimating population abundance or detecting population declines. Care should be taken if using count data during transition months between breeding and non-breeding seasons because our models are based on detection probabilities that may vary in relation to bird behaviour, which changes once breeding commences (Reynolds *et al.* 2015b). Linear regression could be used to validate long-term monitoring indices and evaluate population status of other reintroduced populations that are also marked and systematically monitored. Non-overlap between 95% CIs in any two years indicates a significant difference ($P < 0.05$) in population abundance recorded in those two years, which would serve to alert managers to major changes in the population trajectory, whether it be increasing or in decline.

Acknowledgements

We thank Midway Atoll National Wildlife staff and volunteers for data collection efforts and programmatic support, especially J. Breeden, R. Cope, M. Dalton, D. Dow, M. Duhr-Schultz, W. Holthuijzen, A. Humphrey J. Klavitter, L. Laniawe, A. Munes, B. Ordung and M. Vekasy. We also thank K. Brinck and K. Reynolds for providing useful reviews on a previous version of this paper. Data management from 2008–2010 was provided primarily by L. Laniawe. This study was requested by and primarily funded by Midway Atoll National Wildlife Refuge (including most data collection 2012–2016) and data analysis was supported by U.S. Fish and Wildlife Service's Refuges Inventory and Monitoring Program (IAA 4500036627), U.S. Geological Survey (USGS) Pacific Island Ecosystems Research Center, and USGS Patuxent Wildlife Research Center. Any use of trade, product or firm names is for descriptive purposes only and does not imply endorsement by the U.S. Government.

References

Armstrong, D.P. & Seddon, P.J. 2008. Directions in reintroduction biology. *Trends in Ecology and Evolution* 23: 20–25.

Batson, W.G., Gordon, I.J., Fletcher, D.B. & Mang, A.D. 2015. Review: Translocation tactics: a framework to support the IUCN Guidelines for wildlife translocations and improve the quality of applied methods. *Journal of Applied Ecology* 52: 1598–1607.

Burney, D.A., James, H.F., Burney, L.P., Olson, S.L., Kikuchi, W., Wagner, W.L., Burney, M., McCloskey, D., Kikuchi, D., Grady, F.V., Gage, R. & Nishek, R. 2001. Fossil evidence for a diverse biota from Kaua'i and its transformation since human arrival. *Ecological Monographs* 71: 615–641.

Chapman, D.G. 1951. Some properties of the hypergeometric distribution with applications to zoological sample censuses. *University of California Press* 1: 131–160.

Cooper, A., Rhymer, J., James, H.F., Olson, S.L., McIntosh, C.E., Sorenson, M.D. & Fleischer, R.C. 1996. Ancient DNA and island endemics. *Nature* 381: 484–484.

Executive Order 13022. *Administration of the Midway Islands, October 31, 1996* (61 FR 56875).

Fischer, J. & Lindenmayer, D.B. 2000. An assessment of the published results of animal relocations. *Biological Conservation* 96: 1–11.

International Union for Conservation of Nature (IUCN) 2016. Red List of Threatened Species, version 2016.1. IUCN, Cambridge, UK.

Marshall, A.P. 1992. Censusing Laysan ducks *Anas laysanensis*: a lesson in the pitfalls of estimating threatened species populations. *Bird Conservation International* 2: 239–251.

Miskelly, C.M. & Powlesland, R.G. 2013. Conservation translocations of New Zealand birds, 1863–2012. *Notornis* 60: 3–28.

Moulton, D.W. & Weller, M.W. 1984. Biology and conservation of the Laysan duck. *Condor* 86: 105–117.

Nichols, J.D. 1992. Capture-recapture models. *Bioscience* 42: 94–102.

Olson, S.L. & James, H.F. 1991. Descriptions of thirty-two new species of birds from the Hawaiian Islands: Part I. Non-passeriformes. *Ornithological Monographs* 45: 1–88.

Olson, S.L. & Ziegler, A.C. 1995. Remains of land birds from Lisianski Island, with observations on the terrestrial avifauna of the Northwestern Hawaiian Islands. *Pacific Science* 49: 111–125.

Parker, K.A., Ewen, J.G., Seddon, P.J. & Armstrong, D.P. 2013. Post-release monitoring of bird translocations: why is it important and how do we do it? *Notornis* 60: 85–92.

Presidential Proclamation 8031. 15 June 2006. *Establishment of the Northwestern Hawaiian Islands Marine National Monument* (71 FR 36443).

Reynolds, M.H. & Citta, J.J. 2007. Post-fledging survival of Laysan ducks. *Journal of Wildlife Management* 71: 383–388.

Reynolds, M.H., Slotterback, J.W. & Walters, J.R. 2006. Diet composition and terrestrial prey selection of the Laysan teal on Laysan Island. *Atoll Research Bulletin* 543: 181–199.

Reynolds, M.H., Crampton, L.H. & Vekasy, M.S. 2007. Laysan Teal *Anas laysanensis* nesting phenology and site characteristics on Laysan Island. *Wildfowl* 57: 54–67.

Reynolds, M.H., Seavy, N.E, Vekasy, M.S, Klavitter, J.L. & Laniawe, L.P. 2008. Translocation and early post-release demography of endangered Laysan teal. *Animal Conservation* 11: 160–168.

Reynolds, M.H., Breeden, J.H., Vekasy, M.S. & Ellis, T.M. 2009. Long-term pair bonds in the Laysan duck. *Wilson Journal of Ornithology* 121: 187–190.

Reynolds, M.H., Brinck, K.W. & Laniawe, L. 2011. *Population Estimates and Monitoring Guidelines for Endangered Laysan Teal,* Anas laysanensis, *at Midway Atoll: Pilot Study Results 2008–2010*. Hawai'i Cooperative Studies Unit Technical Report HCSU-021, University of Hawai'i at Hilo, Hilo, Hawaii, USA.

Reynolds, M.H., Courtot, K.N., Berkowitz, P., Storlazzi, C.D. & Flint, E. 2015a. Will the effects of sea-level rise create ecological traps for Pacific island seabirds? *PLOS ONE* 10(9):e0136773.

Reynolds, M.H., Courtot, K.N, Brinck, K.W, Rehkemper, C.L. & Hatfield, J.S. 2015b. Long-term monitoring of endangered Laysan ducks: index validation and population estimates 1998–2012. *Journal of Fish and Wildlife Management* 6: 305–317.

Robson, D.S. & Regier, H.A. 1964. Sample size in Petersen mark-recapture experiments. *Transactions of the American Fisheries Society* 93: 215–226.

SAS Institute Inc. 2012. *SAS Statistical Software v. 9.4*. Cary, North Carolina, USA.

Seavy, N.E., Reynolds, M.H., Link, W.A. & Hatfield, J.S. 2009. Postcatastrophe population dynamics and density dependence of an endemic island duck. *Journal of Wildlife Management* 73: 414–418.

Seddon, P.J., Armstrong, D.P. & Maloney, R.F. 2007. Developing the science of reintroduction biology. *Conservation Biology* 21: 303–312.

Sokal, R.R. & Rohlf, F.J. 1995. *Biometry*. W.H. Freeman, New York, USA.

USFWS (U.S. Fish and Wildlife Service). 2009. Revised recovery plan for the Laysan duck (*Anas laysanensis*). U.S. Fish and Wildlife Service, Portland, Oregon, USA.

Warner, R. 1963. Recent history and ecology of the Laysan duck. *Condor* 65: 1–23.

White, G.C. 2005. Correcting wildlife counts using detection probabilities. *Wildlife Research* 32: 211–216.

Photograph: Laysan Teal with Hawaiian Monk Seal *Neomonachus schauinslandi* on Laysan Island, by Matthew Chauvin.

Reassessing the conservation outlook for Madagascar's endemic Anatidae following the creation of new protected areas

FELIX RAZAFINDRAJAO[1], ANDREW J. BAMFORD[2]*,
H. GLYN YOUNG[3], ARISTIDE ANDRIANARIMISA[4],
ABDALLAH IAHIA BIN ABOUDOU[1] & RICHARD E. LEWIS[1]

[1]Durrell Wildlife Conservation Trust, BP 8511, Antananarivo 101, Madagascar.
[2]Wildfowl & Wetlands Trust, Slimbridge, GL2 7BT, U.K.
[3]Durrell Wildlife Conservation Trust, Les Augrès Manor, Jersey, JE3 5BP, U.K.
[4]Department of Zoology and Biodiversity, Faculty of Science, University of Antananarivo, BP 906, Antananarivo 101, Madagascar.
*Correspondence author. E-mail: Andrew.Bamford@wwt.org.uk

Abstract

Madagascar has three endemic species of Anatidae, all of which are classified by the International Union for Conservation of Nature (IUCN) as Endangered or Critically Endangered. Until recently there have been no protected areas within their ranges to secure key habitat. The creation of several new protected areas in Madagascar since 2010 has created an opportunity for better conservation management of these species, most obviously for Madagascar Pochard *Aythya innotata* which occurs at just a single site that has now been protected. We created distribution models for the other two species, Madagascar Teal *Anas bernieri* and Meller's Duck *A. melleri*, using survey data collected from 2004–2013 and MaxEnt software. Predicted ranges were compared with the locations of protected areas. Additionally, for each species, population monitoring was carried out at one site at which there has been conservation intervention. Our models predicted that breeding Madagascar Teal would occur near healthy mangroves (family: Rhizophoraceae) in areas with high mean temperature, but the total extent of predicted suitable habitat is just 820 km². Non-breeding Meller's Duck favour water surrounded by dense vegetation, in areas with low human population density. Meller's Duck occurs in at least nine protected areas, but most of these were set up for forest conservation and may not support many individuals. Since 2010, two wetland protected areas that could benefit Meller's Duck have been created, although one is small and the other, Alaotra, is heavily disturbed. The population of Meller's Duck at Alaotra is stable. Four new protected areas will benefit Madagascar Teal, covering more than half of the predicted breeding range for this species. The population at one of these protected areas, the

Manambolomaty delta, is increasing. Overall, we conclude that the conservation outlook for Madagascar Teal is improving, but the small range for this species means it is dependent on good management at protected areas where it does occur. Meller's Duck requires more attention, and the outlook for this species remains poor.

Key words: *Anas bernieri*, *Anas melleri*, Madagascar, population trends, species distribution model.

The protected areas created in Madagascar's colonial period and during the first republic (1896–1972) covered only 1.9% of the country's surface area and were gazetted mainly for forest conservation. This left Madagascar's wetlands largely without legal protection (Nicholl & Langrand 1989). In 2003, Madagascar announced an aim to triple the size of its protected area network (known as the Durban Vision; Norris 2006). This new policy provided an opportunity to establish wetland and marine protected areas, with a view to advancing the conservation of endemic and threatened aquatic species. Madagascar's wetlands are in general in very poor condition (Bamford *et al.* 2017) as a result of transformation of wetlands into rice fields, siltation and changes in water quality caused by deforestation and erosion, the presence of invasive non-native species, and over exploitation of resources including mangroves (family: Rhizophoraceae) for wood (Langrand & Goodman 1995; Young 1996a). Given this heavy destruction of Madagascar's wetlands, a study of the distribution, range and population trends for threatened endemic species was considered crucial in order to identify how well served threatened species are by the newly protected areas. Site surveys can provide some answers to these questions

(*e.g.* Young *et al.* 2014), but this approach is not feasible for a country-wide perspective. Species distribution modelling may provide a solution.

Madagascar has three endemic species of Anatidae (Young *et al.* 2013a). Madagascar Pochard *Aythya innotata* are not discussed here as the species occurs at only one site (Bamford *et al.* 2015), making distribution modelling unnecessary. The site at which the pochard occurs has now been protected. The other two species, Madagascar Teal *Anas bernieri* and Meller's Duck *A. melleri*, have different ranges and occur in different habitat types, but both are classified as Endangered by the International Union for Conservation of Nature (IUCN 2016) because they are thought to be suffering long-term population declines. There are few data available for either species, and the evidence for this decline mostly comes from the rate of habitat destruction observed. Historically, neither species had the benefit of protected areas within their range to secure key habitat. The causes of their declines may be similar however (Young *et al.* 2013a), with both being threatened by habitat loss, hunting and fisheries bycatch mortality, and human disturbance (Wilmé 1994; Young *et al.* 2013a,b).

Madagascar Teal is found in the west coastal wetlands of Madagascar, from the

Wildfowl (2017) 67: 72–86

far north of the island (at *c*. 12°S) to 23°S near the town of Toliary. It breeds in mature Black Mangrove *Avicennia marina* and spends the rest of the year on freshwater or brackish lakes (Young 2013; Young *et al.* 2013b). The species is never found more than a few kilometres from the coast, or at altitudes of more than a few metres above sea level (Young 2013). The teal was rediscovered in 1969 (Salvan 1970; Andriamampianina 1976), having been only rarely recorded anywhere in the country for nearly a century. Soon after, 120 individuals were recorded at lakes in the mid-western region of the country (Scott & Lubbock 1974). No more surveys were conducted until the early 1990s, by which time this species was known only in the central-western region (Langand 1990). Further populations have been discovered since then (*e.g.* Razafindrajao *et al.* 2001), but the overall population is estimated to number < 1,700 mature individuals (IUCN 2016).

Meller's Duck is found in wetlands in the highlands of central, east and northwest Madagascar, where this highly territorial species breeds predominantly in forested streams and rivers or in extensive marshland (Young 2013). It was introduced to Mauritius in around 1850 (Morris & Hawkins 1998) but the Mauritian population is now extinct (Young & Rhymer 1988). While there is hardly any information on which to determine population trends for Meller's Duck in Madagascar, numbers are thought to have been declining for the past four to five decades (Young 1996a; Young *et al.* 2013a). Although the species' range covers approximately half of Madagascar, and within that range it occurs on small

rivers and wetlands which provide potentially a very large number of suitable sites, it is rarely seen in groups larger than single figures. The total population size is estimated at < 3,300 birds (IUCN 2016).

This paper aims to assess the current status of these two species, in order to help develop conservation strategies for them. Specific objectives are: 1) to describe trends in numbers at sites where there has been conservation intervention, 2) to utilise survey data to identify suitable habitat and predict the species' distribution, and 3) to compare the predicted distribution with the locations of the newly-protected areas.

Methods

Study sites

Monitoring of the Madagascar Teal population was carried out in the wetlands of the Manambolomaty Delta (18.96°S, 44.35°E) which cover an area of 630 km² (Fig. 1). The delta contains several different types of wetland habitats including freshwater ponds and lakes, estuaries, mangroves and marshes, all of which are utilised by Madagascar Teal, and the wetlands also provide important refuges for other aquatic fauna in western Madagascar (Rabearivony *et al.* 2010). In 1998, a conservation project for the teal was established in the Manambolomaty Delta (see Young *et al.* 2013a) with the objective of ensuring the long-term viability of this population by involving local communities in lake management and monitoring activities.

Meller's Duck population monitoring was conducted at Lake Alaotra and its

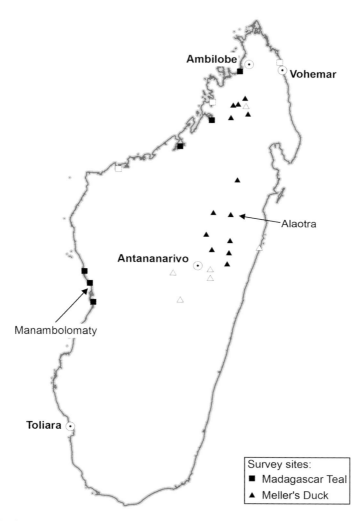

Figure 1. Study sites and other locations referred to in the text. For the survey sites, filled symbols represent sites where the target species was recorded, and unfilled symbols show sites where it was not recorded.

surrounding marshes (17.50°S, 48.50°E) in an area covering 500 km². Lake Alaotra, the largest lake in Madagascar, is surrounded by steep, deforested hills and erosion of the slopes causes substantial sedimentation in the lake. Alaotra is home to several threatened species, including the Alaotra Gentle Lemur *Hapalemur alaotransis* (classed as Critically Endangered by IUCN) which is endemic to the marshes at Alaotra. Alaotra is one of the most threatened ecosystems in Madagascar (Mutschler 2003). Conservation activities at the lake were started in 1995, and include education and awareness programmes, establishing fishermen's groups to regulate activities and monitoring of biodiversity including waterfowl.

In preparation for the distribution surveys, we visited sites throughout the historical range of the two species for collecting count data. For the Madagascar Teal, these ranged from Toliary in the south to Ambilobe in the north, with some sightings also from the northeastern part of the Vohemar district (Young 2013). For Meller's Duck, the historical range covers highland wetlands in the north, east and northwestern regions of the country. Locations of the survey sites are shown in Figure 1.

Distribution surveys

Field work was undertaken from 2002–2006 for Madagascar Teal and 2004–2008 for Meller's Duck, with some extra data collected in 2012 and 2013 for both species. Survey locations were selected opportunistically and informally, sometimes as part of survey work with other objectives, using 1:500,000 scale Foiben-Taosarintan'i Madagascar (FTM) maps and Google Earth images. Most sites were visited only once during the study period. Survey effort varied with the size of the site, ranging from two hours at small ponds to three days at large lakes. The total number of ducks of each species was recorded for each site on each occasion.

Population monitoring

The more detailed monitoring of teal at the Manambolomaty Delta was carried out monthly in at least five months every year from 1999–2012. Counts of birds were made using binoculars and a telescope. Counts commenced between 06:00–12:00 h and the duration of the count was recorded, varying from 15 min to 6 h depending on conditions. A total of 13 locations were monitored during this period, although not all locations were surveyed every year. The number of locations surveyed in each year ranged from 6–13; the number of years in which each location was surveyed ranged from 3 years to all 13 years of the monitoring period.

At Alaotra, biannual monitoring of waterbirds was conducted during July and February (during the dry season and rainy season respectively) from July 1998 to July 2012. In 1998 monitoring was initiated at five locations around the lake, with a sixth site monitored from 2003 onwards. The marsh at one location was converted to rice agriculture in 2001 so monitoring was transferred to a new location nearby. Counts of birds were direct counts using binoculars and telescope. Visits were made by canoe or on foot. Observations began at 05:45 h and ended at 10:00–11:00 h.

Predicted range maps

The survey methods used during the study meant that presence-only modelling was the only appropriate method for distribution modelling. Predicted occurrence maps of Madagascar Teal and Meller's Duck were created using MaxEnt software (Phillips et al. 2006). Both species utilise slightly different habitat during their breeding and non-breeding seasons, often gathering in larger lakes during the non-breeding season

Wildfowl (2017) 67: 72–86

which coincides with the dry season. Madagascar Teal may be constrained by the availability of both breeding and non-breeding habitat (Young *et al.* 2013a), but the model here is based on sightings from the breeding season only. Meller's Duck is also likely to be limited by the availability of both breeding and non-breeding habitat, but there are few breeding records for this species, meaning that it was not possible to develop a satisfactory model of the birds' distribution across breeding habitat. Only sightings made during the dry (non-breeding) season therefore were included in the model.

Habitat variables were prepared in ArcGIS v10 using the Spatial Analyst extension (ESRI 2012). The variables included in each analysis are shown in Table 1, and were all prepared at 1 km resolution. Variables included were: water cover (lakes and rivers) as determined from remote sensing data; vegetation cover relevant to each species (*i.e.* forest cover for Meller's Duck; mangrove cover for Madagascar Teal); a measure of vegetation thickness (the Normalised Difference Vegetation Index, NDVI); human population density as a measure of disturbance; and basic climate information (Table 1). Models for both species covered the entire country. Presence records were filtered so that the minimum distance between sightings was 1 km to match the resolution of the habitat data. The small number of presence records for both species made further compensation for spatial bias unfeasible (see Merow *et al.* 2013), a common problem with rare species. Models were constrained to use only linear relationships. MaxEnt allows the user to set

a regularisation multiplier to reduce over-fitting of the model, but after a few trials we kept this multiplier set to the default of 1 because higher values led to models that predicted very large areas of species occurrence. The software was set to create response curves for each variable and to perform jack-knife measurement of the importance of each variable. Cross-validation was used to test the models, with the number of replicate models set to ten. The final model presented in each case is the average value of all ten models. Good habitat for each species was defined as areas where the final model prediction was ≥ 60% probability of suitability for the species.

Population trends analysis

Count data for both species were modelled using Generalized Linear Models. Statistical analyses were carried out in R 3.2.3 (R Core Team 2015). We tested log-linear models but the results were unsatisfactory as our data were zero-inflated and over-dispersed. Instead we used negative binomial models using the package MASS (Venables & Ripley 2002), which produced a better fit. For each species, counts were modelled with year, site and duration of the count as explanatory variables. Site was fitted as a categorical variable, Year was fitted as a continuous variable to establish an overall trend and then as a categorical variable. Duration was standardised to have mean = 0 and s.d. = 1. The trend for year was plotted with the other variables held stable. The 95% confidence intervals for the predictions were calculated from the standard errors of the predictions.

Table 1. Habitat variables included in MaxEnt models for Madagascar Teal and Meller's Duck.

Variable	Description	Madagascar Teal	Meller's Duck
Wetland area (proportion)	Spatial Analyst software was used to calculate flow accumulation from Digital Elevation Model (DEM) data (Jarvis *et al.* 2008). This was used to create a feature set of rivers and streams. This was combined with the LandSat Global Inland Water dataset (Feng *et al.* 2015) to create an overall water cover feature. Both datasets from 2000.	×	×
Population density (people km^{-2})	Taken from CIESIN *et al.* (2004). Data from 2010.	×	×
Mean temperature (°C)	Annual mean temperature (1970–2000), taken from www.WorldClim.org (Hijmans *et al.* 2005).	×	×
Annual precipitation (cm)	Total annual precipitation (1970–2000), taken from www.WorldClim.org (Hijmans *et al.* 2005).	×	×
Dry season NDVI (October) Rainy season NDVI (February)	NDVI long-term mean (2001–2013) taken from FEWS Net, available at www.earlywarning.usgs.gov/fews and resampled to 1km^2 resolution.	×	×
Elevation (m a.s.l.)	Taken from DEM data (Jarvis *et al.* 2008).	×	×
Slope (% incline)	Calculated from DEM data (Jarvis *et al.* 2008) using Spatial Analyst.		×
Forest cover (%)	Calculated from Madagascar vegetation map (Moat & Smith 2007). Data from 2003–2006.		×
Mangrove cover (%)	Calculated using Giri *et al.* (2011). Data from 1997–2000.	×	

Results

Predicted ranges

Spatial filtering to 1 km resolution resulted in 15 presence records for Madagascar Teal and 33 for Meller's Duck. The habitat preference model for Madagascar Teal during the breeding season was most strongly affected by two variables (variable percentage contributions to the model are given in parentheses): positive effects of mangrove cover (76%) and mean annual temperature (22%). There were also weaker contributions to the model of annual precipitation (negative, 1.4%) and dry season NDVI (positive, 0.1%). The remaining variables made no contribution to the model. On a Receiver Operating Characteristics (ROC) curve, the mean Area Under Curve (AUC) was 0.994 ± 0.003, indicating an excellent model performance.

The main variables in the model for Meller's Duck, all with positive effects, were elevation (41%), water cover (30%) and dry season NDVI (13%). There were also smaller, negative, effects for wet season NDVI (4%), human population density (4%) and mean temperature (3%). The model had AUC = 0.956 ± 0.030.

Predicted distribution maps for both species are shown in Figure 2. Prior to 2010, the majority of breeding habitat for Madagascar Teal was unprotected. Since then, four new protected areas that will benefit this species have been established (Mahavavy-Kinkony, Manambolomaty, Menabe-Antimena and Mangoky-Ihotry). However, three significant areas of predicted good habitat remain unprotected: Ankazomborona, Loza Bay and the Mahajamba Delta. Overall, there was 820 km² of predicted suitable habitat (area where the model predicted suitability for the species greater than 0.6), and of this the proportion protected has increased from 3% to 56% since 2011.

Dry season habitat for Meller's Duck occurs within several long-established protected areas, notably Marotandrano Special Reserve, Zahamena National Park, Ranomafana National Park, Andringitra National Park, and Kalambatritra Special Reserve. Five new protected areas have recently been established that could benefit this species, especially in the northern highlands (Bemanevika and the Tsaratanana-Marojejy corridor) and at Lake Alaotra. The model predicts several small, scattered areas of suitable habitat that are currently unprotected but only one large continuous area of unprotected good habitat: the forests east of Lake Tsiazompaniry in Anosibe. Of the 4,800 km² of predicted suitable habitat (model prediction > 0.6), the proportion protected has increased from 9% to 46%.

Population trends

Models of count data were significantly better fits to the data than null models, both for Madagascar Teal (Likelihood Ratio = 1377, d.f. = 115, $P < 0.001$) and for Meller's Duck (LR = 136, d.f. = 33, $P < 0.001$). The population trends by year are shown in Figure 3. For Madagascar Teal, an initial slight decline has been followed by a steady increase since 2003. By 2011, the population was 136% higher than it had been in 1999, although this was followed by a slight decline again in 2012. With fewer data, the

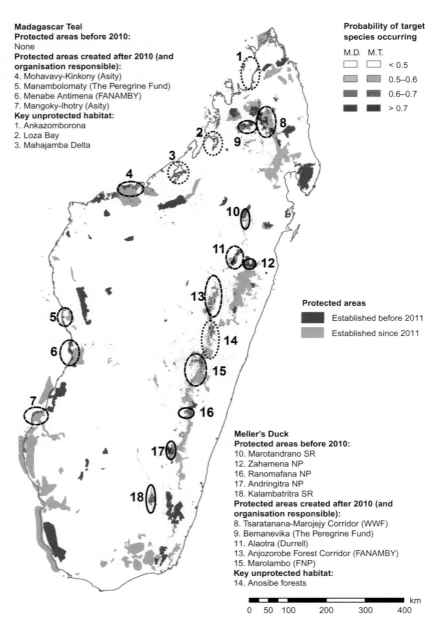

Figure 2. Predicted occurrence of Meller's Duck (blue) and Madagascar Teal (red), with protected areas and key areas of occurrence that are unprotected highlighted. SR = Special Reserve, NP = National Park. "Organisation Responsible" refers to the non-governmental organisation (NGO) that set up the reserve and supports management.

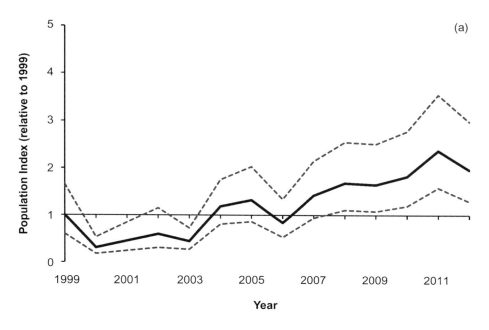

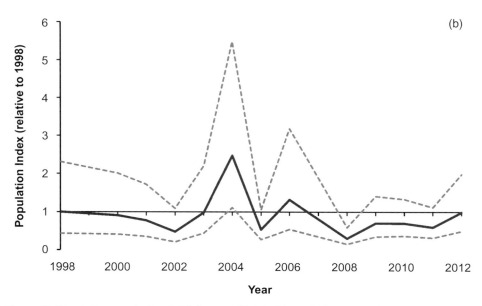

Figure 3. Population trends for a) Madagascar Teal in Manambolomaty, and b) Meller's Duck in Alaotra. Continuous lines = annual population estimate from the model; dashed lines = 95% confidence intervals.

model fit for Meller's Duck had wider confidence intervals and no trend was detectable in the data. The population appears stable based on the limited data available.

Discussion

Our species distribution models broadly confirm previous *ad hoc* distribution maps for these two species (*e.g.* Young 2013), but our maps suggest that there is little suitable habitat within the simplified ranges shown in those *ad hoc* distribution maps. Breeding habitat for Madagascar Teal was predicted to occupy an area of just 820 km^2 along the west coast, constrained nearly entirely by the availability of mangroves. The vegetation map used (Giri *et al.* 2011) generally only includes mature mangroves in its classification (Jones *et al.* 2014), and an effect of dry season NDVI indicates that even among mature mangroves the healthiest stands are being selected (*e.g.* Chellamani *et al.* 2014). Even this small figure may overestimate the suitable habitat, given that Madagascar Teal nest only in Black Mangrove and not in other species, a distinction that the vegetation map does not make. Madagascar Teal is also known to be sensitive to human disturbance (Young *et al.* 2013b), requiring undisturbed wetlands not just for breeding but also for moulting during the dry months.

Dry season habitat for Meller's Duck was associated with wetlands at cooler, high elevations and with high dry season NDVI – generally, areas that are forested or have dense evergreen bush – and in areas with low human population density. We did not record any ducks in the eastern coastal regions, and this was reflected in our model which did not predict any suitable habitat in this region, contrary to previous distribution maps. The eastern wetlands have a high human population density and correspondingly high disturbance, meaning that there is little suitable habitat remaining for Meller's Duck on the east coast (Young 2013). Human disturbance is one of the most important factors affecting freshwater biodiversity in Madagascar (Bamford *et al.* 2017). Apart from the exclusion of the east coast wetlands, the results from our model are similar to previous studies such as Young (1996b) and Langrand (1990), but additionally suggest that the northern highlands may be an important area for this species.

Interpretation of the models needs to be undertaken with caution, as there was no independent dataset available to test the models against. The small sample size limited modelling to linear relationships, but some responses may in reality be non-linear, particularly responses to NDVI. Furthermore, the models cover a fairly short, and already slightly out of date, temporal period and so make no allowance for future changes in habitat associated with climate change or other habitat changes. The effects of climate change, particularly on coastal mangroves, are poorly understood in Madagascar (Ward *et al.* 2016) and may cause substantial changes within protected areas, rendering them unsuitable for some of the species that live there. Despite these limitations, our results do highlight conservation measures that will benefit these species in the short term at least.

Prior to 2010 there were no protected areas within the breeding range of Madagascar Teal, and very few within the dry season range either. Kirindy-Mitea National Park contains two shallow lakes, which shelter approximately a few tens of Madagascar Teal during the dry season, but the park contains no breeding habitat. Since 2010, four protected areas have been created that could benefit breeding Madagascar Teal: Mangoky-Ihotry, Menabe Antimena, Manambolomaty and Mahavavy Kinkony. Between them, these Protected Areas cover more than 50% of the teal's predicted breeding range. Provided they are adequately managed – and in Madagascar that is not a certainty – they represent a major step forward in the conservation of Madagascar Teal.

Meller's Duck occurs in at least five protected areas that were established before 2010 in the highlands. However these areas are primarily for forest protection, and the small rivers and streams that they contain are home to only small numbers of Meller's Duck (Young et al. 2013b). Whilst nearly half of the predicted suitable habitat is now protected, the same problems apply to some of the protected areas created since 2010. Two newly-protected areas may be more useful for Meller's Duck conservation, as they contain more substantial wetland areas: Bemanevika and Alaotra. Bemanevika may contain the most suitable, and least disturbed, habitat, but this site covers only c. 500 ha of wetlands. The Alaotra Protected Area covers a much larger area of wetlands, over 300 km², but this area is highly disturbed (Mutschler 2003; Bamford et al. 2017). If, contrary to our model, the species

does still occur in the east coast wetlands, no new protected areas have been created which would help to reinforce its presence in that region.

Monitoring of Madagascar Teal at Manambolomaty shows that the population is increasing at this site, suggesting that this site at least is adequately managed for this species. While we cannot rule out that this increase may be due to immigration from more disturbed areas, we do know that birds from this site emigrate to other wetlands along the west coast of Madagascar (Razafindrajao et al. 2012). Conservation work at Manambolomaty has worked to make anthropogenic use of the wetlands more sustainable, and our results suggest that this work is effective.

Monitoring of Meller's Duck in Alaotra and its surrounding marshes show that the population is apparently stable over more than ten years of monitoring. This result may mask the seriousness of the situation at Alaotra, as many of the birds recorded there during the dry season will not breed there. The Alaotra wetlands constitute one of the most threatened habitats in Madagascar (Mutschler 2003; Bakoariniaina et al. 2006). Due to intense human pressure, conservation interventions at this site do not appear to be having any beneficial effect for the Meller's Duck. Despite this, Alaotra is still likely home to the largest number of Meller's Ducks in Madagascar, with several hundred birds seen each year, and it is a critical site for this species.

Conclusion

Overall, our results suggest good news for Madagascar Teal. Since 2010 several protected areas have been created, covering more than half of its breeding range, which should, if managed correctly, benefit the species. Monitoring at one of these sites shows a population increase, and monitoring at other sites should be a priority to determine if this increase is occuring more widely. Further detail on nesting habitat requirements for the species is required, although survey work in dense, mature mangroves is not straightforward. However, the total area of breeding habitat is small and non-protected habitat is being cleared rapidly (authors' pers. obs.). Protection of all remaining suitable habitat is in theory achievable and would further improve the outlook for this species, but would require an organisation willing to pay for it – none of the new protected areas in Madagascar have been government funded.

The situation for the Meller's Duck is less promising. The species does occur in several protected areas, and several new ones have been created since 2010 resulting in nearly half of its dry season habitat being protected. However, it does not occur at high density across most of its range. Furthermore, little is known about the species' breeding habitat requirements, so we cannot judge how much protected breeding habitat there might be or how many pairs that could support. There are only two wetland protected areas in its range, one of which is small and the other is large but severely degraded. The non-breeding population at this latter site is stable. This species undoubtedly requires far more conservation attention but first requires more detailed research so that the species' breeding requirements and limiting factors are known in sufficient detail to form a conservation plan.

Acknowledgements

We thank the Durrell teams at Lake Alaotra and Manambolomaty, particularly Jonah Randriamahefasoa, Bary-Jean Rasolonjatovo and Richard Mozavelo and also Robert Bourou and Juliette Velosoa for help with data collection. Geoff Hilton made detailed comments on the manuscript. This paper was made possible by an SCCS Miriam Rothschild Internship awarded to FR in 2015.

References

Andriamampianina, J. 1976. Madagascar. *In* M. Smart (ed.), *Proceedings International Conference on the Conservation of Wetlands and Waterfowl, Heiligenhafen, Federal Republic of Germany, 2–6 December 1974*, pp. 125–126. International Waterfowl Research Bureau, Slimbridge, UK.

Bamford, A.J., Sam, T.S., Razafindrajao, F., Robson, H., Woolaver, L. & René de Roland L.A. 2015. Status and ecology of the last wild population of Madagascar Pochard *Aythya innotata*. *Bird Conservation International* 25: 97–110.

Bamford, A.J., Razafindrajao, F., Young, R.P. & Hilton, G.M. 2017. Profound and pervasive degradation of Madagascar's freshwater wetlands and links with biodiversity. *PLOS ONE* 12: e0182673.

Bakoariniaina, L.N., Kusky, T. & Raharimahefa, T. 2006. Disappearing Lac Alaotra: monitoring catastrophic erosion, waterway silting, and land degradation hazards in Madagascar using LandSat imagery. *Journal of African Earth Sciences* 44: 241–252.

Center for International Earth Science Information Network (CIESIN), International Food Policy Research Institute (IPFRI) & Centro Internacional de Agricultura Tropical

(CIAT) 2004. *Global Rural-Urban Mapping Project (GRUMP): Settlement Points*. Palisades, New York, USA. Available at http://sedac.ciesin.columbia.edu/gpw (last accessed on 26.05.2017).

Chellamani, P., Singh, C.P. & Panigrahy, S. 2014. Assessment of the health status of Indian mangrove ecosystems using multi temporal remote sensing data. *Tropical Ecology* 55: 245–253.

Feng, M., Sexton, J.O., Channan, S. & Townshend, J.R. 2015. Global Inland Water. Global Land Cover Facility, University of Maryland. Available at http://glcf.umd.edu/data/watercover/(last accessed on 26.05.2017).

Giri, C., Ochieng, E., Tieszen, L.L., Zhu, Z., Singh, A., Loveland, T., Masek, J. & Duke, N. 2011. Status and distribution of mangrove forests of the world using earth observation satellite data. *Global Ecology and Biogeography* 20: 154–159.

Hijmans, R.J., Cameron, S.E., Parra, J.L., Jones P.G. & Jarvis, A. 2005. Very high resolution interpolated climate surfaces for global land areas. *International Journal of Climatology* 25: 1965–1978.

International Union for Conservation of Nature (IUCN) 2016. Red List of Threatened Species, version 2016.1. IUCN, Cambridge, UK. Available from www.iucnredlist.org (last accessed 15 August 2017).

Jarvis, A., Reuter, H.I., Nelson, A. & Guevara, E. 2008. *Hole-filled SRTM for the globe Verion 4*. Available from the CGIAR-CSI SRTM 90m database, http://srtm.csi.cgiar.org.

Jones, T.G., Ratsimba, H.R., Ravaoarinorotsihoarana, L., Cripps, G. & Bey, A. 2014. Ecological variability and carbon stock estimates of mangrove ecosystems in north-western Madagascar. *Forests* 5: 177–205.

Langrand, O. 1990. *Guide to the Birds of Madagascar*. Yale University, New Haven, USA.

Langrand, O. & Goodman, S. 1995. Monitoring Madagascar's ecosystems: a look at the past, present and future of its wetlands. *In* T.B. Herman, S. Bondrup-Nielson, J.H. Martin Wilson & N.W.P Munro (eds.), *Ecosystem Monitoring and Protected Areas. Proceedings 2nd International Conference on Science and Management of Protected Areas, Dalhousie University, Halifax, Nova Scotia 16–20 May, 1994*, pp. 204–214. Science and Management of Protected Areas Association, Wolfville, Canada.

Merow, C., Smith, M.J. & Silander, J.A. 2013. A practical guide to MaxEnt for modelling species distributions: what it does, and why inputs and settings matter. *Ecography* 36: 1058–1069.

Moat, J. & Smith, P. 2007. *Madagascar Vegetation Atlas*. Royal Botanical Gardens, Kew, UK.

Morris, P. & Hawkins, F. 1998. *Birds of Madagascar: a Photographic Guide*. Pica Press, Sussex, UK.

Mutschler, T. 2003. Lac Alaotra. *In* S.M. Goodman & J.P. Benstead (eds.), *A Natural History of Madagascar*, pp. 1530–1534. University of Chicago Press, Chicago, USA.

Nicholl, M.E. & Langrand, O. 1989. *Madagascar: Revue de la Conservation et des Aires Protégées*. World Wide Fund for Nature (WWF), Gland, Switzerland.

Norris, S. 2006. Madagascar defiant. *BioScience* 56: 960–965.

Phillips, S.J., Anderson, R.P. & Schapire, R.E. 2006. Maximum entropy modelling of species geographic distributions. *Ecological Modelling* 190: 231–259.

R Core Team 2015. *R: A language and environment for statistical computing*. R Foundation for Statistical Computing, Vienna, Austria. Available from: https://www.R-project.org (last accessed on 31.07.2017).

Rabearivony, J., Thorstrom, R., René de Roland, L.A., Rakotondratsina, M., Andriamalala,

T.R.A., Sam, T.S., Razafimanjato, G., Rakotondravony, D., Raselimanana, A.P. & Rakotoson, M. 2010. Protected area surface extension in Madagascar: do endemisms and threatened species remain useful criteria for site selection? *Madagascar Conservation and Development* 5: 35–47.

Razafindrajao, F., Lewis, R., Nichols, R. & Woolaver, L. 2001. Discovery of a new breeding population of Madagascar Teal *Anas bernieri* in north-west Madagascar. *Dodo* 37: 60–69.

Razafindrajao, F., Young, H.G. & Bin Aboudou, I.A. 2012. Measurements and movements of Madagascar Teal *Anas bernieri* captured and ringed at Lake Antsamaka in central-western Madagascar. *Wildfowl* 62: 165–175.

Salvan, J. 1970. Remarques sur l'évolution de l'avifauna Malgache dupuis 1945. *Alauda* 38: 191–203.

Scott, D. & Lubbock, J. 1974. Preliminary observations on waterfowl in western Madagascar. *Wildfowl* 25: 117–120.

Venables, W.N. & Ripley, B.D. 2002. *Modern Applied Statistics with S, Fourth Edition.* Springer, New York. USA.

Ward, R.D., Friess, D.A., Day, R.H. & MacKenzie, R.A. 2016. Impacts of climate change on mangrove ecosystems: a region by region overview. *Ecosystem Health and Sustainability* 2: e01211.

Wilmé, L. 1994. Status, distribution and conservation of two Madagascar bird species endemic to Lake Alaotra: Delacour's Grebe *Tachybaptus rufolavatus* and Madagascar Pochard *Aythya innotata*. *Biological Conservation* 69: 15–21.

Young, H.G. 1996a. Threatened Anatinae and wetlands of Madagascar: A review and evaluation. *In* M. Birkan, J. van Vessem, P. Havet, J. Madsen, B. Trolliet & M. Moser (eds.), Proceedings of the Anatidae 2000 Conference, Strasbourg, France, 5–9 December 1994. *Gibier Faune Sauvage, Game Wildlife* 13: 801–813.

Young, H.G. 1996b. The distribution and origins of wildfowl (Anatidae) of Western Indian Ocean islands. *In* W.R. Lourenço (ed.), *Biogéographie de Madagascar*, pp. 363–367. Editions de l'Orstom, Paris, France.

Young, H.G. 2013. Family Anatidae. *In* Safford, R.J. & Hawkins, A.F.A. (eds.), *The Birds of Africa. Volume VIII: The Malagasy Region*, pp. 238–262. Christopher Helm, London, UK.

Young, H.G. & Rhymer, J.M. 1988. Meller's Duck: a threatened species receives recognition at last. *Biodiversity and Conservation* 7: 1313–1323.

Young, H.G., Razafindrajao, F. & Lewis, R.E. 2013a. Madagascar's wildfowl (Anatidae) in the new millennium. *Wildfowl* 63: 5–23.

Young, H.G., Razafindrajao, F., Bin Aboudou, A.I., Woolaver, L. & Lewis, R.E. 2013b. Madagascar Teal *Anas bernieri*: a mangrove specialist from Madagascar's west coast. *In* G. Gleason & V.R. Victor (eds.), *Mangrove Ecosystems*, pp. 157–166. Nova Publishers, New York, USA.

Young, H.G., Young, R.P., Lewis, R.E., Razafindrajao, F., Bin Aboudou, I.A. & Fa, J.E. 2014. Patterns of waterbird diversity in central western Madagascar: where are the priority sites for conservation? *Wildfowl* 64: 35–53.

Winter diet of Blue-winged Teal *Anas discors*, Green-winged Teal *Anas carolinensis*, and Northern Shoveler *Anas clypeata* in east-central Texas

DANIEL P. COLLINS[1,2,*], WARREN C. CONWAY[1,3], COREY D. MASON[4] & JEFFREY W. GUNNELS[5]

[1]Arthur Temple College of Forestry and Agriculture, Stephen F. Austin State University, P.O. Box 6109 SFA Station, Nacogdoches, Texas 75962, USA.
[2]Present Address: U.S. Fish & Wildlife Service – Region 2 Migratory Bird Program, P.O. Box 1306, Albuquerque, New Mexico 87103, USA.
[3]Department of Natural Resources Management at Texas Tech University, 15th and Detroit, Mail Stop 42125, Lubbock, Texas 79409, USA.
[4]Texas Parks and Wildlife Department, 4200 Smith School Road, Austin, Texas 78744, USA.
[5]Texas Parks and Wildlife Department, Middle Trinity River Ecosystem Project, 16149 North U.S. Highway 287, Tennessee Colony, Texas, USA.
*Correspondence author. E-mail: Dan_Collins@fws.gov

Abstract

Whilst dabbling duck diet has been studied in some detail on the breeding grounds, it has not been studied as exhaustively at wintering sites in North America. We therefore aimed to describe the diet of Blue-winged Teal *Anas discors*, Green-winged Teal *Anas carolinensis*, and Northern Shoveler *Anas clypeata* using moist-soil managed wetlands in east-central Texas by determining from gut samples the food items ingested and the variation in dry mass of each item taken. A variety of 33 food items (mostly seeds and invertebrates, with only trace amounts of other plant material) were recorded in gut samples of 174 birds. Aggregate dry mass varied among and between species and age-sex cohorts. Several food items occurred frequently in each of the three species – most notably knotweed *Polygonum* sp., panic grass *Panicum* sp., Water-pepper *Persicaria hydropiper* and Curly Dock *Rumex crispus* – indicating the importance of these plant seeds, along with Gastropods, in the diet of dabbling ducks wintering on the wetlands of east-central Texas.

Key words: diet, Blue-winged Teal, food habits, Green-winged Teal, Northern Shoveler, Texas.

Wildfowl (2017) 67: 87–99

A basic knowledge of food requirements, food availability and food preferences is crucial for understanding the ecology of a species (Olney 1963; Dessborn *et al.* 2011). Food intake, which reflects both energetic demands and food resource availability, has a direct influence on body condition and on an individual's ability to survive and breed successfully (Smith & Sheeley 1993). Moreover, for migratory waterbirds, the food ingested and feeding patterns (*e.g.* timing and duration of feeding each day) are likely to vary considerably in relation to the birds' energy needs over the annual cycle and food resources available in different parts of the migratory range (Hartman 1985; Guillemain *et al.* 2007; Arzel *et al.* 2009). While the diet of wintering waterfowl has been relatively under-studied in comparison with studies undertaken during the breeding season, the quality and quantity of food taken plays a key role in influencing not only the birds' winter site selection but their body condition and potentially their survival and productivity (Heitmeyer & Fredrickson 1981; Miller 1986; Euliss & Harris 1987; Moon *et al.* 2007; Callicutt *et al.* 2011).

Dietary studies therefore are one of four key research objectives (along with habitat use, time budgets and body condition) required for a full assessment of the wintering requirements for waterfowl (Korschgen *et al.* 1988). Moreover, many ducks of the *Anatini* tribe share similar feeding behaviours and patterns, including utilising the same food resources during winter, which may lead to interspecific competition or resource partitioning at the wintering sites (DuBowy 1988; Guillemain

et al. 2000). For example, Blue-winged Teal *Anas discors*, Green-winged Teal *Anas carolinensis* and Northern Shoveler *Anas clypeata* all use moderate amounts of semi-aquatic and aquatic vegetation in shallow to moderately deep water habitats (White & James 1978). Northern Shoveler often sieve for small crustaceans in the water column, however, whereas Blue-winged and Green-winged Teal, which seemingly are more generalised in their foraging behaviour and food selection, tend to focus upon plant material (*i.e.* seeds, tubers, or leafy parts of vegetation; Dirschl 1967; Baldassarre & Bolen 1984; Eulis & Harris 1987; Dubowy 1988; Botero & Rusch 1994; Anderson *et al.* 2000; Dessborn *et al.* 2011).

Although feeding behaviour including food selection is typically the product of interactions among biological and nutritional demands, physical capabilities, and environmental conditions (Swanson *et al.* 1974), it tends to vary between species and across seasons, in line with changes in food availability resulting from local environmental conditions. For example, in the case of Blue-winged Teal in the Saskatchewan River Delta, Dirschl (1967) reported fluctuations in the food ingested during summer, when invertebrates dominated the diet, but the seeds of some plant species – sedges *Carex* sp. and *Eleocharis* sp., cattails *Sparganium* sp. and bulrush *Scirpus* sp. – were consistently (18–35% occurrence) consumed over time. Thompson *et al.* (1992) similarly reported that Blue-winged Teal wintering in Mexico consumed > 98% Gastropods during early winter, before switching to a plant-dominated diet (*i.e.* > 96% plant material, primarily stonewort *Chara* sp.) during mid–late winter. Studies of Blue-winged Teal in Central and South America (Botero & Rusch 1994), in Mexico (Saunders & Saunders 1981) and in the southern United States (Swiderek *et al.* 1988) however found that the birds relied primarily on plant food in these areas.

In contrast, Green-winged Teal in the Southern High Plains of Texas were found to consume mostly plant matter and seeds (> 70%; Anderson *et al.* 2000) during autumn and early winter, with considerably less (8–37%) of their diet represented by animal matter (primarily Insecta; Euliss & Harris 1987; Anderson *et al.* 2000), although invertebrates were taken in greater proportion than their availability (Anderson *et al.* 2000). Typically, Green-winged Teal seed consumption reflects the availability of food items in the environment, but they also select larger seeds, such as those produced by *Eleocharis* sp*.,* knotweed *Polygonum* sp*.,* paspalum *Paspalum* sp*.,* barnyard grass *Echinochloa* sp. and docks *Rumex* sp. (Anderson *et al.* 2000). Mouronval *et al.* (2007) also found *Polygonum* to be present in 54% of Eurasian Teal *Anas crecca* collected in northeast France.

Early studies of the Northern Shoveler's diet reported that they primarily consumed vegetation or seeds (Anderson 1959), but more recent work has questioned the extent to which this is prevalent. For example, Mouronval *et al.* (2007) found animal prey in 89% of the samples examined from Northern Shoveler collected in northeast France, whereas although 14 species of seeds were identified, none were frequent. Vest and Conover (2011) found that

Northern Shoveler and Green-winged Teal in a saline system – the Great Salt Lake in Utah – consumed primarily aquatic invertebrates (*i.e.* brine shrimp *Artemia* sp. adults and cysts) during winter, but noted that in late-winter both species increased consumption of wetland plant seeds.

Studies to date therefore suggest that these three dabbling duck species consume similar food items, albeit in differing proportions and that this may also vary with physiological demands, food availability, season and geographic region. The objectives in the study presented here were to quantify and compare the food ingested by Blue-winged Teal, Green-winged Teal, and Northern Shoveler during winter using moist-soil managed wetlands in east-central Texas, to provide information for habitat managers and conservation planners on the food taken by these species when wintering in the region. Whether the sex and age (adult or juvenile) of the birds influenced their diet was also considered. Given that some earlier studies found a seasonal change in the food taken in by the birds, whether there was a change in the birds' diet between autumn and winter, and whether it was evident in one or more of these dabbling duck species wintering in the same area, was investigated in further detail.

Methods

Study area

The study was conducted on the Richland Creek Wildlife Management Area (RCWMA), located in Freestone and Navarro Counties, Texas, USA (31°13'N, 96°11'W). The RCWMA lies almost entirely within the Trinity River floodplain, covering

6,271 ha in the area between the Post Oak Savannah and Blackland Prairie ecological regions (TPWD 2005). All work occurred within the managed moist-soil wetlands of the northern part of the RCWMA.

Sampling gut contents

Birds were collected opportunistically during morning flights or after observation of diurnal foraging (*i.e.* 07:00–18:00 h), from 15 September–28 February in winters 2004/05, 2005/06 and 2006/07 (following Anderson *et al.* 2000). We divided the collection dates into two seasons: 1) autumn (September 1–November 15), and 2) winter (November 16–February 28), to identify any changes in diet between the autumn and mid-winter period. Attempts were made to obtain equal numbers of individuals for each sex within each species, to permit assessment of any differences between the sexes in the food ingested. Upon collection, a 75% ethanol solution was immediately injected into the oesophagus to preserve material *post-mortem* (Anderson *et al.* 2000). Birds were then eviscerated, the digestive tract removed, and stored in 75% ethanol.

Once in the laboratory, digestive tracts (*i.e.* oesophagus, proventriculus and gizzard) were dissected and washed to remove all materials contained within. Birds without any food in their digestive tract were omitted from the analysis. Digestive tract contents were examined under a dissecting microscope (VRW® Stereo Basic Halogen Microscope), animal and plant matter were separated, and these were then identified to lowest taxon possible. Seeds and animal matter were thus identified to genus and species level; other plant material (*e.g.* leaves)

was present in only small amounts and these were recorded as "miscellaneous vegetation" (as in Anderson *et al.* 2000). The food items were dried at 50°C for 24 h, weighed to the nearest 0.1 g, and the percent occurrence and aggregate percent dry mass of each food item was calculated for each bird (following Swanson *et al.* 1974). Aggregate percent dry mass for food taken by each species was determined by summing the dry mass of each food item per individual and converting it into an overall aggregate percent dry mass (Swanson *et al.* 1974; Vest & Conover 2011). All percentage data were arcsine transformed to improve normality (Zar 1999; Vest & Conover 2011).

Data analysis

Multivariate analysis of variance (MANOVA, with Wilks' testing for differences between groups) was used to examine differences in the aggregate percent dry mass of plant and animal food items ingested, among and between species, age-sex cohorts, and between seasons (*i.e.* autumn and winter), following the methods described by Anderson *et al.* (2000) and by Vest and Conover (2011). If significant differences ($P \leq 0.05$) were found between groups in the MANOVA, univariate analysis of variance (ANOVA) was used to identify more specifically whether species, age, sex or season was associated with the food ingested by the birds.

Results

A total of 33 different food items were identified, cumulatively occurring 600 times in all three focal species (Table 1). Plant food items occurred in 98–100% of

Table 1. Percent occurrence and aggregate percent dry mass of food items consumed by Blue-winged Teal (*n* = 66), Green-winged Teal (*n* = 54), and Northern Shoveler (*n* = 54), during winters 2004/05–2006/07 in east-central, Texas. Plants identified to genus or species are all seeds; other vegetation material (*e.g.* leaves) is included as "miscellaneous vegetation".

	Blue-Winged Teal		Green-Winged Teal		Northern Shoveler	
	% Occurrence	% Dry mass	% Occurrence	% Dry mass	% Occurrence	% Dry mass
PLANT	98.4	72.5	100.0	97.5	100.0	93.7
Amaranthus tuberculata	7.6	0.2	11.1	0.4	1.9	0.9
Ammania coccinea	6.1	0.1	1.9	0.0	0.0	0.0
Carex sp.	0.0	0.0	3.7	1.9	5.6	0.4
Chenopodium album	12.1	0.9	16.7	1.5	1.9	0.7
Cyperus erthrorshizos	3.0	0.0	5.6	1.8	1.9	0.0
Cyperus sp.	3.0	0.2	3.7	0.9	3.7	0.5
Echinochloa crusgalli	12.1	1.9	16.7	3.2	18.5	5.8
Echinochloa walteri	7.6	1.9	3.7	0.6	1.9	0.1
Echinodorus rostru	27.3	4.1	18.5	3.2	18.5	4.2
Eclipta prostrate	9.1	1.9	0.0	0.0	1.9	0.3
Eleocharis quadrangulata	13.6	2.4	7.4	1.7	14.8	3.2
Eleocharis sp.	12.1	0.6	14.8	3.4	11.1	3.5
Juncus effusus	0.0	0.0	5.6	0.6	3.7	0.2
Leptochloa fascicularis	3.0	0.1	11.1	2.6	5.6	1.5
Ludwigia peploides	0.0	0.0	0.0	0.0	1.9	0.9
Panicum sp.	47.0	8.1	46.3	12.7	33.3	6.3
Paspalum sp.	6.1	1.2	11.1	4.4	5.6	0.6
Persicaria hydropiper	30.3	10.6	40.7	17.4	25.9	7.4
Polygonum lapathifolia	66.7	26.5	46.3	21.8	48.1	24.3
Polygonum pennsylvanicum	27.3	9.9	18.5	8.8	31.5	20.9
Rumex crispus	37.9	11.2	18.5	7.8	11.1	0.7
Schoenoplectus californicus	10.6	0.8	3.7	1.1	7.4	0.4
Misc. vegetation	9.1	1.3	0.0	0.0	16.7	7.9
ANIMAL	39.4	27.5	7.4	2.5	24.1	6.3
Bivalvia	1.5	0.0	0.0	0.0	0.0	0.0
Corixa sp.	1.5	0.0	0.0	0.0	0.0	0.0
Gastropod pieces	33.3	12.0	1.9	1.1	22.2	9.0
Hermetiaillucens	1.5	0.6	0.0	0.0	0.0	0.0
Hydrophilidae	3.0	0.1	0.0	0.0	0.0	0.0
Odonata	1.5	0.1	1.9	0.3	1.9	1.5
Physidae	3.0	1.2	1.9	0.8	1.9	0.4
Planorbidae	7.6	1.8	0.0	0.0	1.9	0.3
Unidentified invertebrate	1.5	0.0	1.9	0.0	0.0	0.0

individuals, while animal food items were found in 7.4–39.4% of the Blue-winged Teal, Green-winged Teal and Northern Shoveler included in the study (Table 1). On considering seasonal variation in the food taken by each species, plant and animal food items were recorded in 100% and 0–40% respectively for birds collected in autumn, compared with 97–100% plants and 27–41% animal items for the three species during winter (Table 2). The most frequent seed items identified in Blue-winged Teal were Curlytop Knotweed *Polygonum lapathifolium* (67%), panic grass *Panicum* sp. (47%) and Curly Dock *Rumex crispus* (38%), with Gastropod pieces (33%) their most frequently identified invertebrate food. During autumn the most frequent food items in the Blue-winged Teal's diet were Curlytop Knotweed (30%), Gastropod pieces (17%) and Barnyard Grass *Echinochloa crusgalli* (13%), whereas in winter Erect Burhead *Echinodorus rostrus* (22%), California Bulrush *Shoenoplectus californicus* (14%) and Waterhemp *Amaranthus tuberculate* (11%) were more commonly ingested (Table 2). For Green-winged Teal, Curlytop Knotweed (46%), *Panicum* sp. (46%) and Water-pepper *Persicaria hydropiper* (41%) were the most frequently identified seeds overall, and no one invertebrate food item was more frequent than another (Table 1). Seeds ingested by Green-winged Teal were mostly of Curlytop Knotweed (56%), Water Hemp (22%), *Eleocharis* sp. (11%) and Water-pepper (11%) during autumn, and Barnyard Grass (16%), *Eleocharis* sp. (11%) and Erect Burhead (11%) during winter (Table 2). Finally, Curlytop Knotweed (48%), *Panicum* sp. (33%) and Pink Smartweed *Polygonum*

pennsylvanicum (32%) were the most frequent seed items identified in Northern Shoveler throughout the study while Gastropod pieces (22%) were the most frequent invertebrate food item identified (Table 1). In autumn the Northern Shovelers were found to have fed primarily on the seeds of Curlytop Knotweed (33%), Barnyard Grass (16%), Water Pepper (17%) and California Bulrush (17%), and during winter the most frequent food items were Barnyard Grass (19%), Gastropod pieces (15%) and Pink Smartweed (15%) (Table 2).

The aggregate percent dry mass of plant and animal food items varied among species (Wilks' λ = 0.980, d.f. = 2, P = 0.02), but did not vary with season (Wilks' λ = 0.995, d.f. = 1, P = 0.25, n.s.), age (Wilks' λ = 0.997, d.f. = 1, P = 0.43, n.s.) nor sex (Wilks' λ = 0.997, d.f. = 1, P = 0.54, n.s.). There was also no significant interaction between age and sex in the percentage dry mass of plant compared with animal food ingested (Wilks' λ = 0.995, d.f. = 1, P = 0.24, n.s.). Subsequent analysis of variance similarly found that aggregate percent dry mass varied among species (F = 5.57, d.f. = 2, P = 0.003), but was similar for all other comparisons within each species. For example, within Blue-winged Teal, aggregate percent dry mass did not vary between autumn and winter (Wilks' λ = 0.991, d.f. = 1, P = 0.32, n.s.), nor was there any variation between sexes (Wilks' λ = 0.995, d.f. = 1, P = 0.56, n.s.), nor ages (Wilks' λ = 0.994, d.f. = 1, P = 0.46, n.s.), and there was no age x sex interaction (Wilks' λ = 0.989, d.f. = 1, P = 0.26). Similarly, aggregate percent dry mass within Green-winged Teal did not vary between

Table 2. Percent occurrence of food items consumed by Blue-winged Teal, Green-winged Teal, and Northern Shoveler in autumn and winter during 2004/05–2006/07 in east-central Texas. Plants identified to genus or species are all seeds; other vegetation material (*e.g.* leaves) is included as "miscellaneous vegetation".

	Blue-winged Teal		Green-winged Teal		Northern Shoveler	
	Autumn (*n* = 30)	Winter (*n* = 36)	Autumn (*n* = 9)	Winter (*n* = 45)	Autumn (*n* = 6)	Winter (*n* = 48)
PLANT	100.0	97.2	100.0	100.0	100.0	100.0
Amaranthus tuberculata	3.3	11.1	22.2	8.9	0.0	2.1
Echinochloa crusgalli	13.3	8.3	0.0	15.6	16.7	18.8
Shoenoplectus californicus	6.7	13.9	0.0	4.4	16.7	2.1
Carex sp.	0.0	0.0	0.0	2.2	0.0	2.1
Chenopodium album	3.3	2.8	0.0	6.7	0.0	2.1
Cyperus sp.	3.3	2.8	0.0	2.2	0.0	4.2
Eclipta prostrate	3.3	8.3	0.0	0.0	0.0	0.0
Eleocharis sp.	0.0	5.6	11.1	11.1	0.0	8.3
Echinodorus rostru	10.0	22.2	0.0	11.1	0.0	8.3
Juncus effusus	0.0	0.0	0.0	0.0	0.0	0.0
Polygonum lapathifolia	30.0	2.8	55.6	4.4	33.3	12.5
Panicum sp.	3.3	2.8	0.0	8.9	0.0	8.3
Paspalum sp.	0.0	0.0	0.0	2.2	0.0	0.0
Polygonum pennsylvanicum	0.0	5.6	0.0	6.7	0.0	14.6
Cyperus erthrorshizos	0.0	0.0	0.0	0.0	0.0	0.0
Rumex crispus	0.0	0.0	0.0	8.9	0.0	0.0
Leptochloa fascicularis	0.0	0.0	0.0	0.0	0.0	0.0
Eleocharis quadrangulata	0.0	0.0	0.0	0.0	0.0	0.0
Ammania coccinea	0.0	0.0	0.0	0.0	0.0	0.0
Persicaria hydropiper	3.3	2.8	11.1	4.4	16.7	2.1
Ludwigia peploides	0.0	0.0	0.0	0.0	0.0	0.0
Echinochloa walteri	0.0	0.0	0.0	0.0	0.0	0.0
Misc. vegetation	0.0	0.0	0.0	0.0	16.7	0.0
ANIMAL	40.0	41.6	11.1	33.3	0.0	27.0
Bivalvia	0.0	0.0	0.0	0.0	0.0	0.0
Corixa sp.	0.0	0.0	0.0	0.0	0.0	0.0
Gastropod pieces	16.7	11.1	0.0	0.0	0.0	14.6
Hydrophilidae	0.0	0.0	0.0	0.0	0.0	0.0
Odonata	0.0	0.0	0.0	0.0	0.0	0.0
Physidae	0.0	0.0	0.0	0.0	0.0	0.0
Planorbidae	3.3	0.0	0.0	0.0	0.0	0.0
Hermetiaillucens	0.0	0.0	0.0	0.0	0.0	0.0
Unidentified invertebrates	0.0	0.0	0.0	2.2	0.0	0.0

autumn and winter (Wilks' λ = 0.987, d.f. = 1, P = 0.36, n.s.), nor sex (Wilks' λ = 0.986, d.f. = 1, P = 0.33, n.s.), nor age (Wilks' λ = 0.995, d.f. = 1, P = 0.67, n.s.), and there was no age x sex interaction (Wilks' λ = 0.997, d.f. = 1, P = 0.82, n.s.). Aggregate percent dry mass within Northern Shoveler was also remarkably consistent, and did not vary between autumn and winters (Wilks' λ = 0.995, d.f. = 1, P = 0.73, n.s.), nor sex (Wilks' λ = 0.998, d.f. = 1, P = 0.90, n.s.), nor age (Wilks' λ = 0.996, d.f. = 1, P = 0.73, n.s.), and there was no age x sex interaction (Wilks' λ = 0.985, d.f. = 1, P = 0.32, n.s.).

Discussion

Food item occurrence

The results of our study indicated that Blue-winged Teal, Green-winged Teal and Northern Shoveler wintering on the moist-soil managed wetlands in east-central Texas have a varied omnivorous diet. Digestive tract contents consisted mainly of plant seeds, but 39.4%, 7.4% and 24.1% of the food items identified (measured as % occurrence) in the samples collected for each species were invertebrates, predominantly Gastropods. Previous studies of the food taken by two of these species (Green-winged Teal and Northern Shoveler) at different sites but a similar time of year (*i.e.* during migration and in winter) also found that the dabbling ducks ingest a range of food items, albeit that invertebrates comprised a higher proportion of the birds' diet (Vest & Conover 2011; Tietje & Teer 1996). For instance, in a saline system, both Northern Shoveler and Green-winged Teal

foraged primarily on brine shrimp adults and cysts, which accounted for > 70% of the food taken (Vest & Conover 2011). Elsewhere, in southern Texas, Northern Shoveler likewise ingested more animal matter, although this varied with habitat with a higher proportion of animal food found in birds from saline wetlands (where it comprised 80% of the diet) compared with those at freshwater sites (50% of the diet; Tietje & Teer 1996).

Our findings for Green-winged Teal are very similar to those reported by Anderson *et al.* (2000), who found no differences in the aggregate percent dry mass of total seeds or invertebrates consumed by these birds in the Southern High Plains of Texas, 600 km northeast of our own study site. They found that seeds comprised 69–92% and invertebrates 8–31% of the Green-winged Teal's diet, which is generally consistent with the overall occurrence of native seeds (100%) and aquatic invertebrates (7.4%) in the digestive tracts of the individuals collected in the RCWMA. Like Anderson *et al.* (2000), we also found no difference in the occurrence or in the aggregate percent dry mass of invertebrates compared with seeds consumed among the age-sex cohorts for Green-winged Teal. The results concur with those from studies in Europe. Mouronval *et al.* (2007) reported that teal are almost exclusively granivorous in northeastern France, where all samples (n = 48) collected that had food items present contained seeds. Likewise, Arzel *et al.* (2007) found that invertebrate food items were relatively scarce in winter, consistent with studies which show that teal are mainly granivorous during winter not only in

France but also Sweden. Reports on the diet of Northern Shoveler in the United States and Europe during winter generally vary with geographic location. As stated previously, Vest and Conover (2011) found aquatic invertebrates to occur in > 90% of samples, whilst Mouronval *et al.* (2007) found shovelers to be essentially benthivorous, but that they would also frequently consume seeds. Of interest, one plant species – Curlytop Knotweed – was an important and frequent seed identified in shovelers in northeastern France (Mouronval *et al.* 2007), and this was also identified as an important frequent seed species found in Northern Shoveler in the RCWMA.

Variation in food items ingested with habitat and season

That Gastropod pieces were more frequently identified in Blue-winged Teal than in Northern Shoveler at the RCWMA is of interest. Studies of Northern Shoveler elsewhere typically have found a high occurrence of aquatic invertebrates, such as brine fly larvae, brine shrimps, bivalves, Gastropods and Cladocerans in the diet (DuBowy 1988; Thompson *et al.* 1992; Tietje & Teer 1996; Mouronval *et al.* 2007; Vest & Conover 2011), although this seems to vary with geographic location with Gastropods less commonly found in some studies. For example, Northern Shoveler diets consisted primarily of water boatmen *Corixidae* sp. (51.6%), Rotifers (20.4%) and Copepods (15.2%) in California (Euliss *et al.* 1991), whereas in contrast Gastropods were important for the species in southern Texas, where they accounted for 27% and 38% of the food taken by shovelers in freshwater

and saline wetlands respectively during early winter (Tietje & Teer). Thompson *et al.* (1992) found that Gastropods comprised > 98% of the diet of both Blue-winged Teal and Northern Shoveler in Yucatan, Mexico. Gastropod occurrence was not quite as high as this in our own study (~ 26%), but our findings undoubtedly indicate that Gastropods are an important food item for Northern Shoveler in east-central Texas in winter, and for Blue-winged Teal during both autumn and winter (see Tables 1 & 2).

That Gastropods were a major part of the diet for Blue-winged Teal throughout autumn and winter in our study differs from observations made elsewhere, which described seasonal changes in the extent to which Gastropods were taken by the teal, and suggested that the importance of Gastropods also varies with geographic location. In Mexico, for instance, Thompson *et al.* (1992) found that Blue-winged Teal switch from a primarily Gastropod diet during the first half of winter to a seed-based diet during the second half. Moreover, Rollo & Bolen (1969) and Swiderek *et al.* (1988) found that Blue-winged Teal rely primarily on plant foods. In contrast, both seeds and Gastropods were consistently identified as items taken by Blue-winged Teal during both autumn and winter in our study area, which may indicate the availability of these foods for these teal and indeed other dabbling duck species throughout autumn and winter at RCWMA.

Management implications

In this study, native seeds (such as Curlytop Knotweed, Pink Smartweed, Water Pepper, Curly Dock, and *Panicum* sp.) were the most frequent seeds recorded, perhaps due to their hardness and persistence in crops and digestive tracts, or because they were the food most readily available, as suggested by botanical surveys undertaken in the study area (Collins 2012). Botero & Rusch (1994) postulated that *post mortem* digestion of invertebrates may occur and make seeds easier to detect, but the preservation of the gut contents on collecting the birds aimed to keep any such bias to a minimum. Recognising major seasonal foods of importance that influence waterfowl use of areas and how these are obtained through active management practices is key. At moist-soil wetlands such as the RCWMA, water inundation and drawdown regimes are timed precisely in order to promote the germination, growth and seed production of desirable hydrophytic plants (during drawdown), and also to provide food and structural substrates for invertebrate colonisation (during inundation), with both the seeds and invertebrates being key food resources for migrating and wintering waterfowl (Collins *et al.* 2015). Providing suitable habitat for waterfowl should be the main goal of any wetland/waterfowl land manager providing wintering habitat through active management techniques (*e.g.* moist-soil management), as these resources play a key role in improving the probability of successful life history events during other portions of the annual cycle, such as breeding, egg laying, and nesting (Baldassarre *et al.* 1986; Miller 1986; Rave 1987; Thompson & Baldassarre 1990; Rave & Baldassarre 1991; Devries *et al.* 2008; Collins *et al.* 2015). In addition, estimating and collecting long-term data on vegetation

and duck-use days will provide information on potential food production as well as the carrying capacity of the managed wetlands, which can be useful for adjusting management techniques, if necessary, to maximise the birds' use of these sites (Collins 2012). The findings presented here, along with studies conducted concurrently (*i.e.* on aquatic invertebrates and vegetation composition; Collins 2012), will allow area managers to identify potential variables influencing food habits. The data and analyses should provide an understanding of how the dabbling duck species use moist-soil managed wetlands during winter in east-central Texas, and thus inform management schemes (*e.g.* inundation regimes) and influence directly the occurrence of plant species occurrence in these managed moist-soil wetlands.

Acknowledgements

We appreciate thorough reviews by Eileen Rees, Tony Fox, Bruce Dugger and an anonymous referee for useful comments on an earlier version of this manuscript. Financial and logistical support for this research was provided by the Texas Parks and Wildlife Department and the Arthur Temple College of Forestry and Agriculture (MacIntire-Stennis) at Stephen F. Austin State University. We thank Eric Woolverton, Edwin Bowman, Gary Rhodes, Kevin Kraai, and Matt Symmank for field and logistical support.

References

Anderson, H.G. 1959. Food habits of migratory ducks in Illinois. *Illinois Natural History Survey Bulletin* 27: 1–64.

Anderson, J.T., Smith, L.M. & Haukos, D.A. 2000. Food selection and feather molt by nonbreeding American Green-winged Teal in Texas playas. *Journal of Wildlife Management* 64: 222–230.

Arzel, C., Elmberg, J., Guillemain, M., Legagneux, P., Bosca, F., Chambouleyron, M., Lepley, M., Pin, C., Arnaud, A. & Schricke, V. 2007. Average mass of seeds encountered by foraging dabbling duck in Western Europe. *Wildlife Biology* 13: 328–336.

Arzel, C., Elmberg, J., Guillemain, M., Lepley, M., Bosca, F., Legagneux, P. & Nogues, J.B. 2009. A flyway perspective on food resource abundance in a long-distance migrant, the Eurasian teal (*Anas crecca*). *Journal of Ornithology* 150: 61–73.

Baldassarre, G.A. & Bolen, E.G. 1984. Field-feeding ecology of waterfowl wintering on the Southern High Plains of Texas. *Journal of Wildlife Management* 48: 63–71.

Baldassarre, G.A., Whyte, R.J. & Bolen, E.G. 1986. Body condition and carcass composition of nonbreeding Green-winged Teal on the Southern High Plains of Texas. *Journal of Wildlife Management* 50: 420–426.

Botero, J.E. & Rusch, D.H. 1994. Foods of Blue-winged Teal in two Neotropical wetlands. *Journal of Wildlife Management* 58: 561–565.

Callicutt, J.T., Hagy, H.M. & Schummer, M.L. 2011. The food preference paradigm: a review of autumn-winter food use by North American dabbling ducks (1900–2009). *Journal of Fish and Wildlife Management* 2: 29–40.

Chamberlain, J.L. 1959. Gulf Coast marsh vegetation as food of wintering waterfowl. *Journal of Wildlife Management* 23: 97–102.

Collins, D.P. 2012. Moist-soil managed wetlands and their associated vegetative, aquatic invertebrate, and waterfowl communities in east-central Texas. Ph.D. thesis, Stephen F. Austin State University, Nacogdoches, Texas, USA.

Collins, D.P., Conway, W.C., Mason, C.D. & Gunnels, J.W. 2015. Aquatic invertebrate production in moist-soil managed wetlands on Richland Creek Wildlife Management Area, east-central Texas. *The Southwestern Naturalist* 60: 263–272.

Dessborn, L., Brochet, A.L., Elmberg, J., Legagneux, P., Gauthier-Clerc, M. & Guillemain, M. 2011. Geographical and temporal pattern in the diet of pintail *Anas acuta*, wigeon *Anas penelope*, mallard *Anas platyrhynchos* and teal *Anas crecca* in the Western Palearctic. *European Journal of Wildlife Research* 57: 1119–1129.

Deveries, J.H., Brook, R.W., Howerter, D.W. & Anderson, M.G. 2008. Effects of spring body condition and age on reproduction in mallards (*Anas platyrhynchos*). *Auk* 125: 618–628.

Dirschl, H.J. 1967. Foods of Lesser Scaup and Blue-winged Teal in the Saskatchewan River Delta. *Journal of Wildlife Management* 33: 77–87.

DuBowy, P.J. 1988. Waterfowl communities and seasonal environments: temporal variability in interspecific competition. *Ecology* 69: 1439–1453.

Euliss, N.H. Jr. & Harris, S.W. 1987. Feeding ecology of Northern pintails and green-winged teal wintering in California. *Journal of Wildlife Management* 51: 724–732.

Euliss, N.H. Jr., Jarvis, R.L. & Gilmer, D.S. 1991. Feeding ecology of waterfowl wintering on evaporation ponds in California. *Condor* 93: 582–590.

Guillemain, M., Fritz, H. & Guillon, N. 2000. Foraging behavior and habitat choice of wintering Northern Shoveler in a major wintering quarter in France. *Waterbirds* 23: 255–363.

Guillemain, M., Arzel, C., Legagneux, P., Elmberg, J., Fritz, H., Lepley, M., Pin, C., Arnaud, A. & Massez, G. 2007. Teal (*Anas crecca*) adjust foraging depth to predation risk: a flyway-level circum-annual approach. *Animal Behaviour* 73: 845–854.

Hartman, G. 1985. Foods of Mallard, before and during moult, as determined by faecal analysis. *Wildfowl* 36: 65–71.

Heitmeyer, M.E. & Fredrickson, L.H. 1981. Do wetland conditions in the Mississippi Delta hardwoods influence Mallard recruitment? *Transactions of the North American Wildlife Natural Resource Conference* 46: 44–57.

Korschgen, C.E., George, L.S. & Green, W.L. 1988. Feeding ecology of Canvasbacks staging on Pool 7 of the upper Mississippi River. *In* M.W. Weller (ed.), *Waterfowl in Winter*, pp. 237–250. University of Minnesota Press, Minneapolis, USA.

Miller, M.R. 1986. Northern pintail body condition during wet and dry winters in the Sacramento Valley, California. *Journal of Wildlife Management* 50: 189–198.

Miller, M.R. & Eadie, J.M. 2006. The allometric relationship between resting metabolic rate and body mass in wild waterfowl (Anatidae) and an application to estimation of winter habitat requirements. *Condor* 108: 166–177.

Moon, J.A., Haukos, D.A. & Smith, L.M. 2007. Declining body condition of Northern pintails wintering in the Playa Lakes Region. *Journal of Wildlife Management* 71: 218–221.

Mouronval, J.B., Guillemain, M., Canny, A. & Poirier, F. 2007. Diet of non-breeding wildfowl *Anatidae* and Coot *Fulica atra* on the Perhois gravel pits, northeast France. *Wildfowl* 57: 68–97.

Olney, P.J.S. 1963. The food and feeding habits of teal *Anas crecca*. *Journal of Zoology* 140: 169–210.

Rave, D.P. 1987. Time budget and carcass composition of Green-winged Teal wintering in coastal wetlands of Louisiana. M.Sc. thesis, Auburn University, Auburn, Alabama, USA.

Rave, D.P. & Baldassarre, G.A. 1991. Carcass mass and composition of Green-winged Teal wintering in Louisiana and Texas. *Journal of Wildlife Management* 55: 457–461.

Rollo, J.D. & Bolen, E.G. 1969. Ecological relationships of blue and Green-winged Teal on the High Plains of Texas in early fall. *Southwestern Naturalist* 14: 171–188.

Saunders, G.B. & Saunders, D.C. 1981. Waterfowl and their wintering grounds in Mexico, 1937–64. U.S. Fish and Wildlife Service Resource Publication No. 138. USFWS, Washington D.C., USA.

Sheeley, D.G. & Smith, L.M. 1989. Tests of diet and condition bias in hunter-killed Northern Pintails. *Journal of Wildlife Management* 53: 765–769.

Smith, L.M. & Sheeley, D.G. 1993. Factors affecting condition of Northern Pintails wintering in the Southern High Plains. *Journal of Wildlife Management* 57: 226–231.

Smith, T.M. & Smith, R.L. 2006. *Elements of Ecology, 6th Edition.* Pearson Benjamin Cummings, San Francisco, USA.

Swanson, G.A., Meyer, M.I. &. Serie, J.R. 1974. Feeding ecology of breeding blue-winged teals. *Journal of Wildlife Management* 38: 396–407.

Swiderek, P.K., Johnson, A.S., Hale, P.E. & Joyner, R.L. 1988. Production, management, and waterfowl use of sea purslane, Gulf Coast muskgrass, and widgeongrass in brackish impoundments. *In* M.W. Weller (ed.), *Waterfowl in Winter*, pp. 441–457. University of Minnesota Press, Minneapolis, USA.

Texas Parks & Wildlife Department (TPWD). 2005. Texas Comprehensive Wildlife Conservation Strategy 2005–2010. Texas Parks and Wildlife Department, Austin, Texas, USA.

Thompson, J.D. & Baldassarre, G.A. 1990. Carcass composition of nonbreeding Blue-winged Teal and Northern Pintails in Yucatan, Mexico. *Condor* 92: 1057–1065.

Thompson, J.D., Sheffer, B.J. & Baldassarre, G.A. 1992. Food habits of selected dabbling ducks wintering in Yucatan, Mexico. *Journal of Wildlife Management* 56: 740–744.

Tietje, W.D. & Teer, J.G. 1996. Winter feeding ecology of Northern Shovelers on freshwater and saline water wetlands in south Texas. *Journal of Wildlife Management* 60: 843–855.

Vest, J.L. & Conover, M.R. 2011. Food habits of wintering waterfowl on the Great Salt Lake, Utah. *Waterbirds* 34: 40–50.

White, D.H. & James, D. 1978. Differential use of fresh water environments by wintering waterfowl of coastal Texas. *Wilson Bulletin* 90: 99–111.

Zar, J.H. 1999. *Biostatistical Analysis, 4th Edition.* Prentice-Hall, Upper Saddle River, New Jersey, USA.

Photograph: Blue-winged Teal at the Richland Creek Wildlife Management Area by Dan Collins.

Changes in the sex ratio of the Common Pochard *Aythya ferina* in Europe and North Africa

KANE BRIDES[1]*, KEVIN A. WOOD[1], RICHARD D. HEARN[1] & THIJS P.M. FIJEN[2]

[1]Wildfowl & Wetlands Trust, Slimbridge, Gloucestershire GL2 7BT, UK.
[2]Plant Ecology and Nature Conservation Group, Wageningen University & Research, Droevendaalsesteeg 3a, 6708PB Wageningen, the Netherlands.
*Correspondence author. E-mail: Kane.Brides@wwt.org.uk

Abstract

Assessments of the sex ratio among Common Pochard *Aythya ferina* flocks were undertaken in countries across Europe and into North Africa in January 2016, for comparison with results from surveys carried out over the same area in January 1989 and January 1990. The mean (± 95% CI) proportions of males in the population were estimated as 0.617 (0.614–0.620) in 1989–1990 and 0.707 (0.705–0.710) in 2016; this difference between surveys was found to be highly significant. Whilst male bias increased with latitude in both surveys, this relationship was weaker in 2016 as the increases in male bias between 1989–1990 and 2016 were greater in countries further south. Given that the sex ratio of Pochard broods is approximately 1:1 at hatching, the strong male bias observed among adult birds is indicative of lower survival of females compared with males. The results of this study suggest that factors adversely affecting female survival rate (relative to that of males) may partly explain the decline in overall Common Pochard abundance. Given the widespread and ongoing decline of this species throughout most of Europe and North Africa, further information on possible demographic drivers of change is urgently required.

Key words: *Aythya ferina,* Common Pochard, demography, population ecology, sex ratio, species decline.

With an estimated *c.* 600,000 birds in the Central Europe, Black Sea and Mediterranean population and *c.* 250,000 birds in the Northeast/Northwest Europe population, the Common Pochard *Aythya ferina* (hereafter Pochard) is a common and widespread species across Europe and North Africa (Wetlands International

2017). Following steady declines in winter abundance indices since the 1990s (Nagy *et al.* 2014), however, and also a 30–49% decline in breeding abundance over three generations (BirdLife International 2015), its conservation status was up-listed from Least Concern (LC) to Vulnerable (VU) on the European and global IUCN Red Lists in

2015. Given the widespread and ongoing decline of Pochard throughout most of Europe and North Africa, information highlighting possible demographic drivers of the changes in population size therefore is urgently required.

Sex ratio data can potentially provide useful information on the differential survival rates of the sexes (Donald 2007). As such, sex ratio data can potentially be used to infer the demographic causes of declines in population size amongst avian species. In a review of sex ratios of adult birds, Donald (2007) found that increasing male bias is a common feature of threatened populations, and noted that such changes in the sex ratio may reflect differing mortality risks posed to the different sexes. For instance, studies in New Zealand of the Kaka *Nestor meridionalis septentrionalis*, which is included as Endangered (EN) on the IUCN Red List, found that mainland populations exposed to introduced predators have sex ratios that are highly male-biased, whereas populations on predator-free islands have balanced sex ratios (Greene & Fraser 1998). Although there are relatively few cases where the mechanism for higher female mortality has been confirmed, in some migratory species the longer migrations undertaken by the smaller sex (usually female) may put them at greater risk (review in Donald 2007). Moreover, for waterfowl (including Pochard) where the female undertakes most or all of the incubation, nesting females are known to be sensitive to predation during the breeding season (Sargeant *et al.* 1984; Baldassarre & Bolen 1994; Blums *et al.* 1996). These factors can lead to a skew

towards males among adults, which may be particularly evident among breeding birds. It therefore seems that changes in the variables that influence sex ratios can translate into population-level effects, such as declining numbers, and that information on any change in the sex ratio for a species can provide a valuable insight into its population processes (Donald 2007).

Any assessment of changes in sex ratio must, however, account for potential differential spatial patterns in the distributions of males and females. Many duck populations wintering in the northern hemisphere exhibit geographical gradients in sex ratio, with greater proportions of males wintering further north (Bellrose *et al.* 1961; Perdeck & Clason 1983; Owen & Dix 1986). In the case of the Pochard, Owen & Dix (1986) found that the sex ratio among flocks of the species at sites in the United Kingdom during winter 1983/84 was highly correlated with latitude, with a greater male bias in more northerly areas. At a larger spatial scale, analysis of sex ratio data recorded for Pochard across Europe and into North Africa during surveys made in January 1989 and January 1990 similarly found a latitudinal effect, with a higher proportion of males recorded at higher latitudes (Carbone & Owen 1995).

In the study presented here we assess the sex ratios among Pochard wintering across Europe and into North Africa in January 2016, for comparison with those from the 1989–1990 survey over the same area (reported by Carbone & Owen 1995), and consider the results in light of the population decline recorded between the two surveys. Given the observed decline in the number of Pochard recorded over the past 20 years, and the increasing male bias found in declining populations of other species, we hypothesised that the male bias in the Pochard population would be greater in the more recent survey.

Methods

Sex ratio survey

National coordinators of the annual International Waterbird Census (IWC), which has been organised by Wetlands International (previously the International Waterfowl and Wetlands Research Bureau; IWRB) each year since 1967, were asked to organise sex ratio determinations of Pochard to be undertaken by their network of volunteer counters during the mid-January IWC in 2016. Bird-watchers were also invited to submit data collected outside of the IWC counts and the project was heavily promoted on social media, using the Twitter hashtag #Pochard to encourage interest among the bird-watching community.

Observers were asked to record for each site visited the total flock size, the number of birds for which sex was determined, the number of males and the number of females, location name, latitude and habitat. Data were submitted online via the Duck Specialist Group website (http://www.ducksg.org/projects/compoch/), or via various online recording portals used by waterbird counters and birdwatchers, including BirdTrack and Observation.org. In order to ensure that the results were comparable with those of the 1989–1990 survey (*i.e.* Carbone & Owen 1995), only data collected during a 16-day period (9–24

January 2016) were used in the analysis, with 45.5% (n = 834 flocks; 106,288 individual birds) of the sex ratio samples collected on the 2016 IWC focal dates (16–17 January).

For some surveys, sex could not be determined for all individuals within the flock, in which case the sex ratio in the sample was assigned to the full flock to yield weighted estimates of the total numbers of males and females, after Carbone & Owen (1995). Following the approach of earlier studies (*e.g.* Sheldon 1998; Hardy 2002; Donald 2007), sex ratio was expressed as the proportion of males within the sample, calculated as:

$$\text{Sex ratio} = n_m/(n_m + n_f),$$

where n_m and n_f refer to the total numbers of males and females, respectively. This formula allowed the sex ratio to be calculated for any sample of birds, including for an individual country or the total population. Whilst data for individual flocks were not reported in Carbone & Owen (1995), data on the numbers of males and females were presented for each country, which allowed comparison with our 2016 data. Countries for which data were not available for either the 1989–1990 or 2016 surveys were not included in the analysis. Because of the strong relationship between male bias and latitude, standardising the surveyed area to a consistent set of countries was necessary to permit comparison of the 1989–1990 and 2016 survey results. In total, data from 13 countries were available from both the 1989–1990 and 2016 surveys: Algeria, Britain, Denmark, France, Germany,

Greece, Hungary, Italy, the Netherlands, Republic of Ireland, Romania, Spain and Switzerland. For the 1989–1990 and 2016 surveys, we calculated the sex ratio for each country based on the total numbers of males and females in that country, whilst the population sex ratio was based on the total numbers of males and females in all 13 countries.

To verify that the coverage achieved by the 1989–1990 and 2016 surveys were comparable, we estimated the proportion of the total population of Europe and North Africa that were within the countries included in the surveys, based on the mean IWC counts over the four years leading up to the survey years (*i.e.* in January 1985–1988 and January 2012–2015). The IWC mean count data showed that these 13 countries included in both surveys accounted for 0.724 and 0.729 of the total numbers of Pochard in 1985–1988 and 2012–2015, respectively, suggesting that the 1989–1990 and 2016 surveys were based on almost identical proportions of the total populations.

Statistical analyses

All statistical analyses were carried out using R version 3.3.0 (R Development Core Team 2016), with statistically significant results attributed where $P < 0.05$; all P values were adjusted using Holm-Bonferroni corrections to account for multiple comparisons (Holm 1979). First, to assess whether the sex ratio of: (i) each country, and (ii) the total population, differed significantly between the two surveys, we used a 2-sample binomial test for equality of proportions to assess whether the proportion of males in the 1989–1990 survey differed significantly from the proportion of males recorded in 2016 (Crawley 2005). Second, for each survey we used a two-tailed binomial test to assess the significance of the deviation of the total numbers of males and females for: (i) each country, and (ii) the total population, from a 1:1 ratio. In both cases, the binomial tests allowed 95% confidence intervals to be estimated for the proportion of males in the sample, based on the approach of Clopper & Pearson (1934).

For the country-level weighted sex ratio data for both surveys, we used linear models with Gaussian error structures to assess the relationships between the sex ratio (expressed as the proportion of males) and: (i) survey year, (ii) central latitude for each country, and (iii) the interaction between year and latitude. The inclusion of survey year, and its interaction with latitude, allowed us to test whether the proportion of males differed between years, and whether the magnitude of any difference was consistent over the range of latitudes surveyed. Following inspections of the model residuals, to meet the assumptions of the linear modelling approach we square-root transformed the response variable (sex ratio) (Zuur *et al.* 2010).

Results

The mean (± 95% CI) proportion of males in the population was estimated as 0.617 (0.614–0.620) in 1989–1990, and 0.707 (0.705–0.710) in 2016 (Table 1); a 2-sample binomial test for equality of proportions indicated that this 0.09 difference in the sex ratio was highly significant ($\chi^2_1 = 1893.58$,

Table 1. A comparison of the weighted sex ratios (proportion of males) for each country included in the current study and the earlier 1989–1990 survey of Carbone and Owen (1995), as well as for the total populations. The sample sizes (*n*) represent the total numbers of individuals counted in each country. All *P* values were adjusted using Holm-Bonferroni corrections for multiple comparisons.

Country	Latitude (decimal degrees)	1989–90 survey					2016 survey					Comparison between surveys		
		Sex ratio (prop. males)	Lower 95% CI	Upper 95% CI	Binomial test P value	*n*	Sex ratio (prop. males)	Lower 95% CI	Upper 95% CI	Binomial test P value	*n*	Change in prop. males	Test for equal proportions χ^2	P value
Algeria	28.00	0.112	0.092	0.136	<0.001	836	0.458	0.314	0.608	1.000	48	0.346	44.65	<0.001
Britain	53.83	0.716	0.708	0.723	<0.001	14,722	0.713	0.705	0.721	<0.001	12,413	−0.002	0.13	1.000
Denmark	56.00	0.713	0.697	0.729	<0.001	3,153	0.699	0.675	0.723	<0.001	1,397	−0.014	0.84	1.000
France	47.00	0.625	0.618	0.631	<0.001	22,351	0.700	0.677	0.723	<0.001	1,609	0.076	36.54	<0.001
Germany	51.00	0.654	0.641	0.666	<0.001	5,654	0.723	0.717	0.729	<0.001	20,722	0.069	102.14	<0.001
Greece	39.00	0.476	0.470	0.482	<0.001	25,702	0.566	0.554	0.577	<0.001	7,735	0.089	189.99	<0.001
Hungary	47.00	0.858	0.797	0.906	<0.001	176	0.617	0.582	0.650	<0.001	811	−0.241	36.47	<0.001
Italy	43.00	0.616	0.590	0.642	<0.001	1,409	0.685	0.678	0.693	<0.001	15,569	0.069	28.29	<0.001
Netherlands	52.32	0.653	0.643	0.662	<0.001	9,971	0.739	0.732	0.747	<0.001	12,949	0.086	200.52	<0.001
Republic of Ireland	53.00	0.750	0.717	0.782	<0.001	705	0.747	0.688	0.803	<0.001	231	−0.003	0.00	1.000
Romania	46.00	0.754	0.747	0.762	<0.001	13,310	0.666	0.640	0.691	<0.001	1,361	−0.089	50.93	<0.001
Spain	40.00	0.522	0.507	0.538	0.029	4,285	0.682	0.671	0.693	<0.001	7,016	0.160	287.00	<0.001
Switzerland	46.83	0.676	0.637	0.714	<0.001	591	0.747	0.741	0.752	<0.001	24,527	0.071	14.47	0.002
Total	–	**0.617**	**0.614**	**0.620**	**<0.001**	**102,865**	**0.707**	**0.705**	**0.710**	**<0.001**	**106,388**	**0.090**	**1893.58**	**<0.001**

$P < 0.001$). The numbers of males were significantly higher than expected for a 1:1 ratio for every country in the 1989–1990 survey except Algeria and Greece (which had significant female biases), and for every country in the 2016 survey except Algeria (which had no significant bias) (Table 1). Of the 13 countries compared between 1989–1990 and 2016, eight showed significantly greater male bias in 2016, whilst two showed significantly reduced male bias (Table 1; Fig. 1). However, these two countries, Hungary and Romania, contributed relatively few

birds (811 and 1,361, respectively) to the total sample, and hence did not counteract the overall pattern of greater male bias in 2016 that was observed for the total population.

We found significant positive effects of both latitude and year on the proportions of males in each country ($F_{3,22} = 18.65$, $P < 0.001$, $R^2_{adj} = 67.9\%$; Table 2; Fig. 2). Furthermore, we found evidence of a significant, negative interaction between latitude and year, such that the increases in male bias in 2016 were greater for countries

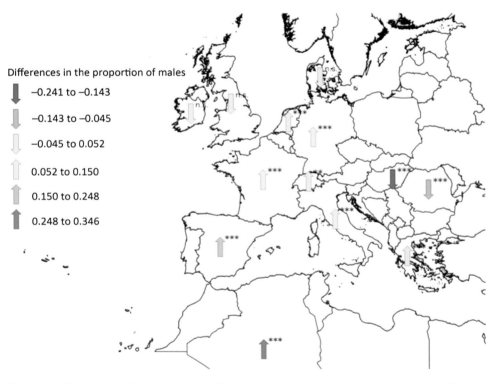

Differences in the proportion of males

- −0.241 to −0.143
- −0.143 to −0.045
- −0.045 to 0.052
- 0.052 to 0.150
- 0.150 to 0.248
- 0.248 to 0.346

Figure 1. The size and direction of the difference in male bias in the sex ratio (expressed as the proportion of males) between the 2016 and 1989–1990 surveys. Positive values indicate a greater male bias in 2016 compared with 1989–1990. The statistical significance of the difference is indicated for each country (see Table 1): *** = $P < 0.001$, ** = $P < 0.01$, n.s. = $P > 0.05$.

Table 2. The mean (± s.e.) estimates and significance of the effect sizes of the parameters in our linear model on square-root-transformed sex ratio (proportion of males) in the 13 countries surveyed in both 1989–1990 and 2016. All *P* values were adjusted using Holm-Bonferroni corrections for multiple comparisons and are statistically significant.

Parameter	Estimate	s.e.	*t* value	*P* value
Intercept	−38.3006	11.6570	−3.29	0.029
Latitude	0.7781	0.2483	3.13	0.034
Year	0.0193	0.0058	3.31	0.029
Latitude * Year	−0.0004	0.0001	−3.09	0.034

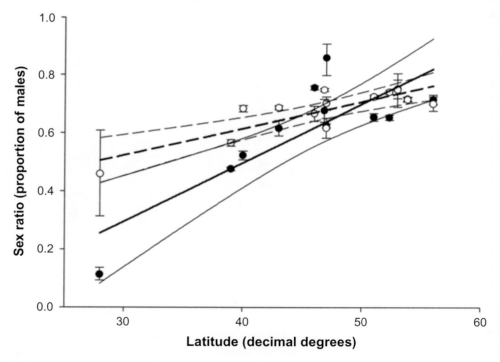

Figure 2. The mean (± 95% CI) effect of latitude on Pochard sex ratio, based on the 1989–1990 (solid circles and lines) and 2016 (open circles and dashed lines) surveys.

at lower latitudes, albeit sample sizes for Algeria in 2016 were low (Table 2; Fig. 2).

Discussion

The 2016 survey found, as expected, that Pochard sex ratios continue to show a strong male bias, as previously described by Owen & Dix (1986) and Carbone & Owen (1995), with a relatively greater proportion of females wintering in countries in the south of the range. We also found evidence that this latitudinal effect on sex ratio was weaker in 2016 than in 1989–1990, however, with lower latitude countries showing the greatest increases in male bias. Males are typically dominant over females in this species, and therefore are able to occupy more favourable wintering areas that are closer to the breeding grounds, resulting in the sub-dominant females moving further afield (Choudhury & Black 1991). The greater capacity of males to withstand cold temperatures has also been proposed as an explanation for the tendency observed in small-bodied birds, including ducks, for a higher proportion of males to winter at higher latitudes (Ketterson & Nolan 1976; Nichols & Haramis 1980; Owen & Dix 1986; Carbone & Owen 1995; Evans & Day 2001). For example, Owen & Dix (1986) calculated that the lower critical temperature for Pochard (representing the minimum air temperature at which an animal can maintain its basal metabolic rate whilst resting without incurring additional thermoregulatory costs) was 7.1°C for males but 8.4°C for females.

The analyses also indicated that the male bias for all Pochard surveyed across Europe and North Africa was significantly greater in 2016 (0.71 males in all birds surveyed) than in 1989–1990 (0.62 males). The two surveys covered the same 13 countries and an almost identical proportion of the population, suggesting that the observed change in sex ratio represents a real change in population structure, rather than being an artefact of the sampling process. Given that we had data only for two survey periods, it was not possible to determine whether the difference in sex ratios formed part of a trend of increased male bias over time. The change in the proportion of males recorded in each of the different countries, and the continued relationship between sex ratio and latitude, does however reinforce the view that there was a greater proportion of males in the population in 2016. Countries at lower latitudes were more likely to show an increased male bias in the sex ratio in 2016 compared with 1989–1990, whereas more northerly countries typically showed little or no change. For instance, we found no difference in the proportion of males reported in Denmark between study periods, which concurs with Christensen & Fox (2014) who similarly found no significant trend in the sex ratio of Pochard wintering in Denmark between 1982 and 2010 on analysing hunter-shot wing samples. Nonetheless, that age checks for Pochard in low-latitude countries such as Algeria, Greece, Italy and Spain, and also across the whole study area, all showed stronger male bias in 2016 may have consequences for Pochard population dynamics and for the success of any conservation measures. For example, Donald (2007) noted that increasingly male-biased sex ratios can lead to lower *per capita* productivity where intense male competition hinders female reproduction. Given the current skew, productivity in Pochard is likely to be limited not by males but by the number of adult breeding females, as found in other duck populations (*e.g.* Hoekman *et al.* 2002).

Given that the sex ratio of Pochard broods is approximately 1:1 at hatching (Blums & Mednis 1996), the male bias observed among full-grown birds suggests a lower survival of females compared with males. Moreover, the declining population size together with the greater male bias recorded in 2016 indicates that female survival rates may have decreased more sharply than those of males. Theoretically male survival rates could have increased more than those for females, but given the overall population trend this is unlikely. The observed change in the sex ratio could alternatively have resulted from a shift in winter distribution, resulting in a lower proportion of all females being counted during the IWC, but this seems also unlikely for two key reasons: (1) the proportion of the total numbers that were included in the two sex ratio surveys were almost identical, and (2) where shifts in duck distribution have been found in Europe they have typically been to the north and east, which given the positive relationship between the proportion of males and latitude should have resulted in more females and fewer males being counted in 2016, which was not the case. Shifts in winter distribution, generally northeast towards breeding areas, have been demonstrated for a wide range of species, including some diving ducks, in recent decades (Lehikoinen *et al.* 2013). For Pochard, declines in wintering numbers in the west of the range have tended to be greater than the overall population trend (*e.g.* 65% decline between winters 1988/89 and 2013/14 in the UK, Hayhow *et al.* 2017; *c.* 60% during 2004–2013 in the Netherlands, Hornman *et al.* 2015) and increases have been observed further east (*e.g.* wintering numbers have increased in Sweden, Nilsson 2008). However, our finding that the 13 countries included in both sex ratio surveys held a consistent proportion of 0.72–0.73 of the total number of Pochard counted by the IWC suggests that any redistribution of birds from the westernmost countries was largely within the area covered by this survey and thus did not affect the overall sex ratio assessment.

There are several potential direct and indirect factors which may explain the apparent relative decrease in female survival. The two most likely are changes in levels of: (i) direct and indirect hunting mortality, and (ii) predation. Pochard is a widely huntable species, being legal quarry in at least 26 countries throughout Europe (Powolny & Czajkowski, in press). Waterbird hunting pressure within Europe is widely recognised as being greatest in southern Europe (Mateo 2009) where the largest proportion of female Pochard overwinter, a view supported by a recent compilation of hunting bag estimates for 17 European countries by Powolny & Czajkowski (in press). In addition, Pochard are huntable from 1 August in several east European countries, which could cause additional selective hunting mortality on breeding females and naïve juveniles prior to their departure to winter quarters, with most males having already migrated away by the beginning of August (Fox *et al.* 2016). Spatial differences in hunting pressure may also influence female survival indirectly through mortality on the birds ingesting spent lead gunshot. The prevalence of lead gunshot ingestion by waterbirds varies considerably among European countries but tends to be higher in southern Europe (Mateo 2009). Furthermore, two recent studies have demonstrated a relatively high susceptibility of Pochard to lead ingestion, both within the UK (Green & Pain 2016) and across Europe (Andreotti *et al.* in press). Whilst the latter did not investigate sex differences in Pochard mortality attributable to lead poisoning, the authors estimated that it accounts for the death of 56,511 Pochard in Europe each year (Andreotti *et al.* in press). Given the apparent elevated risk to female Pochard from direct and indirect hunting related mortality, greater hunting pressure in southern Europe therefore could be an important factor in the apparent decrease in the proportion of females in the population.

Increased female mortality from predation is another possible reason underlying the shift in the sex ratios between surveys, with a recent review of the decline in Pochard numbers suggesting that predation by a range of species (including non-native mammals), could be a major threat to breeding Pochard populations (Fox *et al.* 2016). Female Pochard spend more time on the breeding grounds, incubating eggs and raising ducklings, than do males, leading to an increased level of predation, particularly during the incubation period. Devineau *et al.* (2010) reported that for Green-winged Teal *Anas crecca* lower female survival rates compared with males were apparent during the breeding season, likely resulting from greater predation of females, but not during winter. Such nest predation is often by non-native mammal species, such as Racoon Dog *Nyctereutes procyonoides* and, particularly, American Mink *Mustela vison* (*e.g.* Blums *et al.* 1996). Although data on the population trends of invasive non-native mammals are scarce, it is widely recognised that they are increasing, especially Raccoon Dog which now occurs in 21 countries in Europe (Genovesi *et al.* 2009).

The causes of the changes in sex ratio in time and space remain unknown and further research to estimate and compare male and female survival rates (*e.g.* using a capture-mark-resight approach; White & Burnham 1999) would provide valuable demographic information needed to help understand these changes. Such an analysis of marked birds would allow spatial and temporal patterns in survival to be assessed, and potential drivers of survival rates to be examined (*e.g.* Wood *et al.* in press). Furthermore, such an approach can identify whether changes in survival are limited to particular age classes, sexes, or regions, all of which are currently unknown for Pochard. Previous research has estimated survival rates for female Pochard breeding in Latvia (*e.g.* Blums *et al.* 1996, 2002), but we currently lack comparable survival estimates for males. Collecting information on hunting bag sizes, and other sources of hunting mortality, could permit assessment of the relative impact on male *versus* female Pochard. It would also be advantageous to initiate routine collection of sex ratio data as part of standard waterbird surveys, such as the IWC. Such data would help to provide a better understanding of the patterns of inter-annual variation in sex ratios, and thus differentiate between long-term trends and fluctuations between years. Annual data collection is ongoing in some European countries, but currently too few to describe population-scale trends. Surveys of duck wings from hunted birds can also provide a valuable additional source of data for assessing trends in the sex and age composition of the population (Christensen & Fox 2014).

Overall, we currently lack the knowledge to explain the demographic causes and underlying drivers of the observed decrease in the proportion of female Pochard wintering in Europe and North Africa. Nevertheless, our findings suggest that female survival in relation to that of males is now lower than during the late 1980s, and that aspects of the life-cycle more strongly pertinent to females are likely to be contributing to the observed decline in overall population size. Better monitoring and analyses of demographic information will help to elucidate the situation and to facilitate the further development of targeted conservation and management actions for this declining duck species.

Acknowledgements

We thank the national coordinators and the many volunteers across Europe and North Africa who collected, submitted and collated the data upon which our analyses are based. We also thank Szabolcs Nagy and Tom Langendoen at Wetlands International for the provision of IWC data. Our thanks go to Nick Moran and Neil Calbrade for the provision of counts submitted to BirdTrack and the Wetland Bird Survey, Johannes Wahl and Nicolas Strebel for supplying data from Ornitho, to Hisko de Vries for data from Observation.org and Timme Nyegaard for the provision of data from Dofbasen. Jonathan Cooper kindly helped with the sorting of count data. Thibaut Powolny (OMPO) and Matt Ellis (BASC) kindly provided the European hunting bag estimates. We are grateful to Eileen Rees, Tony Fox, Geoff Hilton and Matthieu Guillemain for their helpful comments on an earlier version of our manuscript. Finally, our grateful thanks go to Matthieu Guillemain for his input at the project planning stage and everyone at the Pochard discussion at the 4th Pan-European Duck Symposium in Finland for their thoughts and useful discussions on data needs for Pochard.

References

Andreotti, A., Guberti, V., Nardelli, R., Pirrello, S., Serra, L., Volponi, S. & Green, R.E. In press. Economic assessment of wild bird mortality induced by the use of lead gunshot in European wetlands. *Science of the Total Environment* doi: 10.1016/j.scitotenv.2017.06.085.

Baldassarre, G.A. & Bolen, E.G. 1994. *Waterfowl Ecology and Management*. John Wiley & Sons, Inc., New York, USA.

Bellrose, F.C., Scott, T.G., Hawkins, A.S. & Low, J.B. 1961. Sex ratios and age ratios in North American ducks. *Illinois Natural History Survey Bulletin* 27: 391–471.

BirdLife International. 2015. *European Red List of Birds*. Office for Official Publications of the European Communities, Luxembourg.

Blums, P. & Mednis, A. 1996. Secondary sex ratio in Anatinae. *Auk* 113: 505–511.

Blums, P., Mednis, A., Bauga, I., Nichols, J.D. & Hines, J. E. 1996. Age-specific survival and philopatry in three species of European ducks: a long-term study. *Condor* 98: 61–74.

Blums, P., Nichols, J.D., Hines, J.E. & Mednis, A. 2002. Sources of variation in survival and breeding site fidelity in three species of European ducks. *Journal of Animal Ecology* 71: 438–450.

Carbone, C. & Owen, M. 1995. Differential migration of the sexes of Pochard *Aythya ferina*: results from a European survey. *Wildfowl* 46: 99–108.

Choudhury, S. & Black, J.M. 1991. Testing the behavioural dominance and dispersal hypothesis in Pochard. *Ornis Scandinavica* 22: 155–159.

Christensen, T.K. & Fox, A.D. 2014. Changes in age and sex ratios amongst samples of hunter-shot wings from common duck species in Denmark 1982–2010. *European Journal of Wildlife Research* 60: 303–312.

Clopper, C.J. & Pearson, E.S. 1934. The use of confidence or fiducial limits illustrated in the case of the binomial. *Biometrika* 26: 404–413.

Crawley, M.J. 2005. *Statistics: An Introduction Using R.* John Wiley & Sons, Chichester, UK.

Devineau, O., Guillemain, M., Johnson, A.R. & Lebreton, J.D. 2010. A comparison of Green-winged Teal *Anas crecca* survival and harvest between Europe and North America. *Wildlife Biology* 16: 12–24.

Donald, P.F. 2007. Adult sex ratios in wild bird populations. *Ibis* 149: 671–692.

Evans, D.M. & Day, K.R. 2001. Migration patterns and sex ratios of diving ducks wintering in Northern Ireland with specific reference to Lough Neagh. *Ringing & Migration* 20: 358–363.

Fox, A.D., Caizergues, A., Banik, M.V., Devos, K., Dvorak, M., Ellermaa, M., Folliot, B.,

Green, A.J., Grüneberg, C., Guillemain, M., Håland, A., Hornman, M., Keller, V., Koshelev, A.I., Kostiushyn, V.A., Kozulin, A., Ławicki, Ł., L.uigujõe, L., Müller, C., Musil, P., Musilová, Z., Nilsson, L., Mischenko, A., Pöysä, H., Ščiban, M., Sjeničić, J., Stīpniece, A., Švažas, S., Wahl, J. 2016. Recent changes in the abundance of Common Pochard *Aythya ferina* breeding in Europe. *Wildfowl* 66: 22–40.

Genovesi, P., Barcher, S., Kobalt, M., Pascal, M. & Scalera, R. 2009. Alien Mammals of Europe. *In* J.A. Drake (ed.), *Handbook of Alien Species in Europe Invading Nature: Springer Series Invasion Ecology.* Springer, Dordrecht, the Netherlands.

Green, R.E. & Pain, D.J. 2016. Possible effects of ingested lead gunshot on populations of ducks wintering in the UK. *Ibis* 158: 699–710.

Greene, T.C. & Fraser, J.R. (1998). Sex ratio of North Island kaka (*Nestor meridionalis septentrionalis*), Waihaha Ecological Area, Pureora Forest Park. *New Zealand Journal of Ecology* 22: 11–16.

Hardy, I.C.W. (ed.) 2002. *Sex Ratios: Concepts and Research Methods.* Cambridge University Press, Cambridge, UK.

Hayhow, D.B., Bond, A.L., Douse, A., Eaton, M.A., Frost, T., Grice, P.V., Hall, C., Harris, S.J., Havery, S., Hearn, R.D., Noble, D.G., Oppel, S., Williams, J., Win, I. & Wotton, S. 2017. *The State of the UK's Birds 2016.* The RSPB, BTO, WWT, DAERA, JNCC, NE, NRW and SNH. Sandy, Bedfordshire, UK.

Hoekman, S.T., Mills, L.S., Howerter, D.W., Devries, J.H. & Ball, I.J. 2002. Sensitivity analyses of the life cycle of midcontinent mallards. *Journal of Wildlife Management* 66: 883–900.

Holm, S. 1979. A simple sequentially rejective multiple test procedure. *Scandinavian Journal of Statistics* 6: 65–70.

Hornman, M., Hustings, F., Koffijberg, K., Klaassen, O., van Winden, E., Sovon Ganzen- en Zwanenwerkgroep & Soldaat, L. 2015. *Watervogels in Nederland in 2013/2014*. Sovon rapport 2015, RWS-rapport BM 15.21. Sovon Vogelonderzoek Nederland, Nijmegen, the Netherlands.

Ketterson, E.D. & Nolan, V. 1976. Geographic variation and its climatic correlates in the sex ratio of eastern wintering Dark Eyed Juncos (*Junco hyemalis hyemalis*). *Ecology* 57: 679–693.

Lehikoinen, A., Jaatinen, K., Vähätolo, A.V., Crowe, O., Deceuninck, B., Hearn R.D., Holt, C.A., Hornman, M., Keller, V., Nilsson, L. Langendoen, T., Tománková, I., Wahl, J. & Fox, A.D. 2013. Rapid climate driven shifts in wintering distributions of three common waterbird species. *Global Change Biology* 19: 2071–2081.

Mateo, R. 2009. Lead poisoning in wild birds in Europe and the regulations adopted by different countries. *In* R.T. Waston, M. Fuller, M. Pokras & W.G. Hunt (eds.), *Ingestion of Lead from Spent Ammunition: Implications for Wildlife and Humans*, pp. 71–98. The Peregrine Fund, Boise, USA.

Nagy, S., Flink, S. & Langendoen, T. 2014. *Waterbird Trends 1988–2012: Results of Trend Analyses of Data from the International Waterbird Census in the African-Eurasian Flyway*. Wetlands International, Ede, the Netherlands.

Nichols, J.D. & Haramis, G.M. 1980. Sex-specific differences in winter distribution patterns of Canvasbacks. *Condor* 82: 406–416.

Nilsson, L. 2008. Changes in numbers and distribution of wintering wildfowl in Sweden. *Ornis Svecica* 18: 135–226.

Owen, M. & Dix, M. 1986. Sex ratios in some common British wintering ducks. *Wildfowl* 37: 104–112.

Perdeck, A. C. & Clason. C. 1983. Sexual differences in migration and wintering ducks ringed in the Netherlands. *Wildfowl* 34: 137–143.

Powolny, T. & Czajkowski, A. (eds). In press. *The Conservation and Management of Selected Huntable Bird Species in Europe. A Review of Status of the Annex 2A Bird Directive Species*. OMPO Publishing, Vilnius, Lithuania.

R Development Core Team. 2016. *R: A Language and Environment for Statistical Computing. Version 3.3.0*. R Foundation for Statistical Computing, Vienna, Austria. URL http://www.R-project.org/.

Sargeant, A.B., Allen, S.H. & Eberhardt, R.T. 1984. Red Fox predation on breeding ducks in mid-continent North America. *Wildlife Monographs* 89: 1–41.

Sheldon, B.C. 1998. Recent studies of avian sex ratios. *Heredity* 80: 397–402.

Wetlands International. 2017. *Waterbird Population Estimates*. Wetlands International, Wageningen, The Netherlands. Available from wpe. wetlands.org (last accessed 24th August 2017).

White, G.C. & Burnham, K.P. 1999. Program MARK: survival estimation from populations of marked animals. *Bird Study* 46: S120–S139.

Wood, K.A., Nuijten, R.J.M., Newth, J.L., Haitjema, T., Vangeluwe, D., Ioannidis, P., Harrison, A.L., MacKenzie, C., Hilton, G.M., Nolet, B.A. & Rees, E.C. In press. Apparent survival of an Arctic-breeding migratory bird over 44 years of fluctuating population size. *Ibis* doi: 10.1111/ibi.12521.

Zuur, A.F., Ieno, E.N. & Elphick, C.S. 2010. A protocol for data exploration to avoid common statistical problems. *Methods in Ecology & Evolution* 1: 3–14.

Facultative heterospecific brood parasitism among the clutches and broods of duck species breeding in South Bohemia, Czech Republic

PETR MUSIL*, ZUZANA MUSILOVÁ & KLÁRA POLÁKOVÁ

Department of Ecology, Faculty of Environmental Sciences,
Czech University of Life Sciences, Kamýcká 1176, CZ-165 21, Prague 6, Czech Republic.
*Correspondence author. E-mail: p.musil@post.cz

Abstract

Heterospecific brood parasitism (HBP) frequently occurs in waterfowl, though much less often than conspecific brood parasitism. In this study, we assess the rate of HBP among clutches and broods of five sympatric breeding duck species: Gadwall *Anas strepera*, Mallard *Anas platyrhynchos*, Red-crested Pochard *Netta rufina*, Common Pochard *Aythya ferina* and Tufted Duck *Aythya fuligula* from nest and brood surveys carried out in the Třeboň Biosphere Reserve and surrounding area (South Bohemia, Czech Republic) in 2006–2015 inclusive. Assessment of 2,323 clutches and 3,056 broods found a higher rate of HBP in clutches than in broods. The rate of HBP in the broods of host birds did not increase with the rate of HBP in host clutches for the five species investigated. The highest proportion of brood parasitism recorded was among Red-crested Pochard. Tufted Duck showed the lowest difference in the HBP rate between clutches and broods; Mallard the highest. From the parasitising female's perspective, the rate of HBP in clutches increased with the rate of HBP in broods for each species investigated. We can conclude that the choice of host affects the success of HBP (*i.e.* the frequency of HBP in clutches *vs.* rate of HBP in broods), and that this can differ between the five species included in the study. Tufted Duck seems to be the most suitable host species as well as the most successful parasite.

Key words: ducks, parasitised broods, parasitised nests.

Facultative brood parasitism is an alternative but not necessarily exclusive reproductive strategy where a parasitising female lays its eggs in another bird's nest then leaves the host to incubate the eggs and raise the hatchlings. This strategy is common in many precocial birds, including waterfowl (Weller 1959; Payne 1977; Yom-Tov 1980;

Rohwer & Freeman 1989), probably due to the lower costs of brood care for the parasitised female compared to those accruing to females of altricial species (Davies 2000). There are obvious benefits to the parasitising female that can lead to this behaviour. For instance, female fitness may be enhanced without incurring the energetic costs of incubation and brood-rearing (Yom-Tov 1980; Sorenson 1998; Andersson & Åhlund 2000), although the increase in clutch size can potentially reduce hatching success (Davies & Baggott 1989; Sayler 1992; Amat 1993; Sorenson 1997; Kear 2005). In spite of this, the strategy is common in ducks (Sayler 1992; Geffen & Yom-Tov 2001), probably because the costs to both the parasitising and the parasitised female is relatively low compared to those in atricial species (Lyon & Eadie 1991; Sorenson 1992; Deeming 2002). Females can lay eggs in the nest of the same species (conspecific brood parasitism – CBP) or, in the nests of other species (heterospecific brood parasitism – HBP), as reviewed by Kear (2005). HBP occurs in all waterfowl groups and in all geographic regions (Yamauchi 1995; Geffen & Yom-Tov 2001; Kear 2005; Krakauer & Kimball 2009). However, in contrast to CBP, HBP is much less frequently observed and studied.

We have been monitoring HBP at a site in the Czech Republic since 1999. Our long-term study follows similar work conducted in the same area during the 1970s (Smrček 1981). In an earlier analysis, Musil & Neužilová (2009) found that HBP occurred in 6.6% of nests monitored in South Bohemia, Czech Republic in 1999–2007. This rate was lower than that recorded across almost the same area in 1975–1980, a period when the breeding population size of most duck species peaked in South Bohemia, and 13.9% of clutches were found to have been parasitised (Smrček 1981).

The present study aims to compare recent (2006–2015) data collected on the rate of HBP found in clutches with HBP rates recorded for broods for five sympatric breeding duck species: Gadwall *Anas strepera*, Mallard *Anas platyrhynchos*, Red-crested Pochard *Netta rufina*, Common Pochard *Aythya ferina* and Tufted Duck *Aythya fuligula*. The presence of parasitic nestlings in broods is considered visual evidence of successful HBP. Given that Musil and Neužilová (2009) found inter-specific differences in the probability of HBP being recorded within clutches, we predicted that there would similarly be statistically significant variation between the five species in the rate of HBP recorded in broods, and that the difference in the HBP rate recorded in broods *versus* in HPB clutches (indicative of the success of the HPB strategy and the suitability of the host species for being parasitised) would also vary across the five species being investigated. We hypothesised that the Red-crested Pochard, found in earlier studies to be the species with both the highest rate of HBP in its nests and the most frequent parasite (Musil & Neužilová 2009), would be the most suitable host species as well as the most successful parasite, with the lowest reduction in the proportion of HBP recorded in broods compared to clutches.

Methods

Nest surveys

Nest surveys were made on islands and in the littoral stands of fishponds in the Třeboň Biosphere Reserve and surrounding area in South Bohemia, Czech Republic (48.97–49.26°N, 14.66–14.97°E) from 2006 to 2015 inclusive. Nest survey sites were the same as those covered by subsequent brood counts (see below). Each pond was visited at 7–14 day intervals from May to July, and each nest was checked at least twice during incubation. The occurrence of HBP was determined by the different colour, size and shape of the eggs (Weller 1959; Amat 1991; Dugger & Blums 2001; Šťastný & Hudec 2016). Heterospecific clutch parasitism was defined as any nest that contained at least one egg of a different species.

Brood counts

Brood counts were carried out on the same wetlands as nest surveys from April to August. For each brood sighted, the species of the adult female, the age of brood (Gollop & Marshall 1954), and the number of ducklings of each species were recorded (*e.g.* Šťastný & Hudec 2016). It was assumed that broods containing ducklings of a different species represented HBP. While such instances could be the result of post-hatch brood amalgamation or accidental mixing, data from monitoring of individually-marked females at our site (*i.e.* marked by nasal saddles) suggest that post-hatch mixing virtually never occurred. No case of post-hatch brood mixing was documented for 189 marked females on the study area between 2006 and 2015, but we recorded 23 broods reared by marked females that contained one or more ducklings of other species. We considered there to be a record of HBP if at least one duckling of a different species was found in the brood of the host female.

Data analysis

Data from the nest surveys and from the brood counts were used to calculate the percentage of incubated clutches and reared broods of a given species that had been parasitised by another species (*i.e.* where the clutch or brood contained at least one egg or duckling of a different duck species). Additionally, the extent to which a given species parasitised other species was calculated as the number of occasions on which the species' eggs were found in a clutch being incubated by a different species divided by the total number of clutches containing the eggs of that species (*i.e.* the sum of the number of parasitised clutches and the number of the species' own clutches), with brood data being treated in the same way.

The difference in the proportion of HBP recorded in clutches and in broods for each species in each year was calculated to assess the success of HPB in relation to both the parasitising and the parasitised species. Multiple linear regression (in Statistica version 13) was then used to analyse the effects of year and species on this difference (arcsine transformed) between clutches and broods in the levels of parasitism recorded.

Results

In total, 2,323 clutches and 3,056 broods were recorded in South Bohemia between 2006 and 2015. Among these, HBP was found in 228 clutches (9.8%) and 133 broods (3.7%). The highest frequency of HBP was found in both the clutches and broods of Red-crested Pochard. The frequency of HBP was higher in clutches than in broods for all species, but the difference was least pronounced in Tufted Duck (Table 1). There was no significant

Wildfowl (2017) 67: 113–122

Table 1. Occurrence of heterospecific brood parasitism among the nests and broods of five species of ducks breeding at a site in South Bohemia, Czech Republic, 2006–2015.

Species	Clutches		Broods	
	No. incubated	% parasitised (*n*)	No. reared	% parasitised (*n*)
Mallard	711	9.7% (69)	922	2.3% (21)
Gadwall	274	13.5% (37)	687	2.3% (16)
Red-crested Pochard	30	40.0% (12)	133	8.3% (11)
Common Pochard	583	8.4% (49)	702	2.7% (19)
Tufted Duck	725	8.4% (61)	612	7.5% (46)
Total	2,323	9.8% (228)	3,056	3.7% (113)

Table 2. Parasitism rates expressed as the percentage of duck nests or broods found to have been parasitised by a given species in South Bohemia, Czech Republic. The values show the number of occasions when eggs or ducklings were found in a clutch or brood incubated/reared by a different species in relation to the total number of clutches or broods with at least one egg or duckling of that species (*i.e.* parasitised nests plus the species own clutches or broods). * = sum of the species' own clutches and the number of clutches where it was found to have parasitised another species. ** = sum of the species' own broods and the number of broods where it was found to have parasitised another species.

Species	Clutches		Broods	
	No. containing at least one egg of this species*	% cases where these were parasitising another species (*n*)	No. containing at least one duckling of this species**	% cases where these were parasitising another species (*n*)
Mallard	728	2.3% (17)	932	1.1% (10)
Gadwall	284	3.5% (10)	695	1.2% (8)
Red-crested Pochard	55	45.5% (25)	172	22.7% (39)
Common Pochard	641	9.0% (58)	742	5.5% (41)
Tufted Duck	759	4.5% (34)	635	3.3% (21)
Total	2,467	5.8% (144)	3,175	3.7% (119)

correlation between frequency of HBP recorded in the clutches and broods of the five species considered (Spearman rank correlation: $r_s = 0.44$, $n = 5$, $P = 0.46$, n.s.).

Red-crested Pochard was more likely than the other species considered to exhibit parasitism; 25 (45.5%) of 55 clutches and 39 (22.7%) of 172 broods with at least one Red-crested Pochard egg or duckling were cases of HBP (Table 2). Conversely, only 4.5% of 759 clutches and 3.3% of 635 broods that included Tufted Duck eggs or ducklings were being reared by another species (Table 2). The proportion of parasitism by a given species in clutches and in broods was highly correlated (Spearman rank correlation: $r_s = 1.00$, $n = 5$, $P \leq 0.02$;

Fig. 1). Although this relationship was driven mainly by Red-crested Pochard, which appeared to have a relatively high parasitising rate at both the clutch and brood-rearing stage, this correlation remained significant on excluding Red-crested Pochard from the analysis (Spearman rank correlation: $r_s = 1.00$, $n = 4$, $P \leq 0.05$; Fig. 1).

The degree of reduction in the rate of HBP in broods compared to clutches varied significantly between species (Fig. 2, Table 3). This change was markedly lower in Tufted Duck than in the other species, both for the probability of the birds being parasitised by another duck species and for the probability of Tufted Duck parasitising

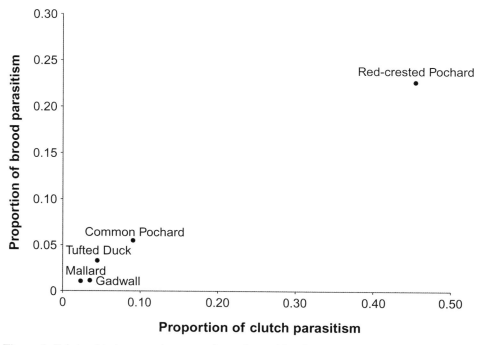

Figure 1. Relationship between the proportions of parasitism by a given species in clutches and in broods.

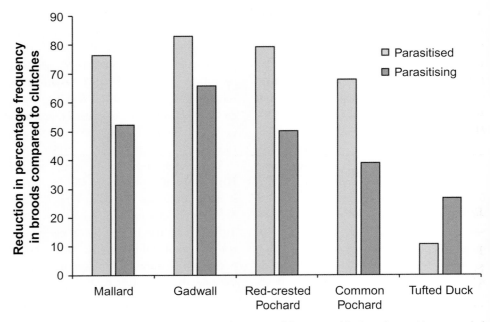

Figure 2. Reduction in the frequency (% occurrence) of heterospecific brood parasitism recorded during the breeding season (*i.e.* the decrease in HBP between clutches and broods), both when parasitising other species and on being parasitised by another species.

others (Fig. 2). There was no evidence to suggest that the HBP rate varied significantly between years during the 2006–2015 study (Table 3).

Discussion

In general, the level of breeding parasitism was found to be higher in clutches than in broods, probably because of disadvantages to the parasitic eggs and ducklings experienced during incubation and early brood rearing. Previous studies have found that, compared to non-parasitic eggs, parasitic eggs suffer more from possible nest desertion of large clutches, egg breakage, false timing of egg-laying, differential post-hatch survival and associated lower survival of large broods, and incomplete imprinting (Andersson & Eriksson 1982; Geffen & Yom-Tov 2001; Birkhead & Brillard 2007). The species in the present study appear to differ in the extent to which they are used as hosts and also in the success of the HBP ducklings in their host broods. Tufted Duck seem to be the species for which parasitic females benefit most during incubation and early brood rearing, with 8.4% of HBP in clutches compared to 7.5% of HBP in broods. Conversely, the proportion of parasite ducklings in Mallard, Gadwall and Red-crested Pochard broods compared to their occurrence in nests of dabbling duck species was very low.

Table 3. Effects of species and year on the difference (reduction) in the frequency of heterospecific brood parasitism (HBP) recorded in broods compared with the frequency of HBP recorded in clutches in South Bohemia, Czech Republic (GLM analysis). (a) = probability of given species being parasitised by another species; (b) = probability a given species parasitising another species (measured as the number of occasions on which the eggs/ducklings of a given species were found in the clutch/brood of a different species, in relation to the total number of all clutches/broods containing eggs/duckling of the given species); n.s. = not significant.

(a) Probability of being parasitised by other duck species				
Effect	Estimate	F	d.f.	P
Intercept	1.716	15.52	1	0.001
Species	1.346	3.04	4	0.036
Year	0.923	0.93	9	n.s.

(b) Probability of parasitising other duck species				
Effect	Estimate	F	d.f.	P
Intercept	3.892	50.157	1	0.001
Species	1.510	4.864	4	0.005
Year	1.298	1.860	9	n.s.

Of the five duck species considered, Red-crested Pochard most frequently parasitised other species, and was also the species most frequent parasitised. The higher rate of breeding parasitism in Red-crested Pochard than in other European duck species has been recorded in several other studies across its breeding range, with HBP noted both in clutches (Amat 1987, 1991, 1993; Fouzari *et al.* 2015) and in broods (Keller 2014). The findings are also in accordance with a comparison of the rate of breeding parasitism recorded in clutches for the same duck species in the same study area over two earlier time periods (1975–1980 and 1999–2007; see Musil & Neužilová 2009). However, the assumption that species parasitising at a higher rate of HBP and presumably benefiting from this alternative reproductive strategy (Lyon & Eadie 1991; Sorenson 1992) have higher success (*e.g.* measured as having a similar HBP ratio at the brood-rearing stage as at the egg-laying stage) was not confirmed. The level of nest parasitism recorded in clutches of Tufted Duck and Common Pochard in the study area agrees with that reported by others (Mednis 1968; Bezzel 1969; Mlíkovský & Buřič 1983). There is still very little data on parasitism rates recorded for the broods of other European duck species, however, for comparison with the results presented here.

The occurrence of brood parasitism is determined by the costs and benefits of the parasitic behaviour for a parasitic female (Sorenson 1992). Some females lay parasitic eggs before they start their own nests, potentially enhancing their reproductive success in this way (Åhlund & Andersson 2001). The main benefit of parasitism in these cases is the ability to reproduce without caring for the eggs and hatchlings, ultimately increasing female productivity. When parasitic eggs are laid by females during the build-up to the main breeding season, before establishing their own nest site, we can therefore expect that these females are in good condition (Owen & Black 1990; Kear 2005). Females may also lay parasitic eggs to invest and obtain reproductive success when they have failed to compete successfully for nest sites, however, or when they lost their own nest due to predation or bad weather conditions (*i.e.* they make the "best of bad job": Payne 1977; Yom-Tov 1980; Davies 2000); such females may be in poorer condition. These patterns could explain the high success of HBP among Tufted Duck, which showed only little difference between the rate of HBP in clutches and in broods. Because this duck species breeds relatively late in

comparison with the other duck species (Neužilová & Musil 2010; Šťastný & Hudec 2016), only early-breeding females can parasitise or conversely could be parasitised by other species. Generally, in the case of early breeders, they are assumed to be in better condition and a higher success of HBP among these individuals could be expected (Owen & Black 1990; Bowler 2005).

Based on the findings of this study, we conclude that the choice of the host female's nest could affect the subsequent success of HBP, which can be measured as the occurrence of ducklings in parasitised broods.

Acknowledgements

We are very grateful to all the co-workers involved in the breeding waterbird counts in South Bohemia in 2006–2015. Among many others, we would personally like to thank Magda Brožová, Markéta Čehovská, Milan Haas, Tereza Kejzlarová, Blanka Kuklíková, Anna Langrová, Hana Malíková, Michaela Nachtigalová and Šárka Neužilová for their help in the field. Moreover, we are grateful to Steve Ridgill for improving our English. Anthony D. Fox, Bruce Dugger, Eileen Rees and an anonymous reviewer kindly provided valuable comments to earlier versions of the manuscript. We are grateful to Jan Zouhar for advice concerning statistical analyses. This study has been supported by a grant from the Czech University of Life Sciences Prague (IGA FŽP No. 20154260) and by Grant No. EHP-CZ02-OV-1-007-01-2014.

References

Åhlund, M. & Andersson, M. 2001. Female ducks can double their reproduction. *Nature* 414: 600–601.

Amat, J.A. 1987. Is nest parasitism among ducks advantageous to the host? *The American Naturalist* 130: 454–457.

Amat, J.A. 1991. Effects of Red-crested Pochard nest parasitism on Mallards. *Wilson Bulletin* 103: 501–503.

Amat, J.A. 1993. Parasitic laying in Red-crested Pochard *Netta rufina* nests. *Ornis Scandinavica* 24: 65–70.

Andersson, M. & Åhlund, M. 2000. Host-parasite relatedness shown by protein fingerprinting in a brood parasitic bird. *Proceedings of National Academy of Sciences USA* 97: 13188–13193.

Andersson, M. & Eriksson, M.O.G. 1982. Nest parasitism in Goldeneyes *Bucephala clangula* – some evolutionary aspects. *The American Naturalist* 120: 1–16.

Bezzel, E. 1969. *Die Tafelente (Aythya ferina)*. Die neue Brehm-Bücherei, Vol. 405. A. Ziemsen Verlag, Wittenberg-Lutherstadt, Germany.

Birkhead, T.R. & Brillard, J.P. 2007. Reproductive isolation in birds: postcopulatory prezygotic barriers. *Trends in Ecology & Evolution,* 22: 266–272.

Bowler, J. 2005. Breeding strategies and biology. *In* J. Kear (ed.), *Bird Families of the World: Ducks, Geese and Swans*, pp. 65–111. Oxford University Press, Oxford, UK.

Davies, A.K. & Baggott, G.K. 1989. Egg-laying, incubation and intraspecific nest parasitism by the Mandarin Duck *Aix galericulata*. *Bird Study* 36: 115–122.

Davies, N.B. 2000. *Cuckoos, Cowbirds and Other Cheats*. T. & A.D. Poyser, London, UK.

Deeming, D.C. 2002. *Avian Incubation*. Oxford University Press, London, UK.

Dugger, D.B. & Blums, P. 2001. Effect of conspecific brood parasitism on host fitness for Tufted Duck and Common Pochard. *The Auk* 118: 717–726.

Fouzari, A., Samraoui, F., Alfarhan, A.H. & Samraoui, B. 2015. Nesting ecology of Ferruginous Duck *Aythya nyroca* in north-eastern Algeria. *African Zoology* 50: 299–305.

Geffen, E. & Yom-Tov, Y. 2001. Factors affecting the rates of intraspecific nest parasitism among Anseriformes and Galliformes. *Animal Behaviour* 62: 1027–1038.

Gollop, J.B. & Marshall, W.H. 1954. *A Guide for Aging Duck Broods in the Field*. Mississippi Flyway Council Technical Section, Northern Prairie Wildlife Research Center, Mississippi, USA. Version 14 November 1997 accessible online at http://www.npwrc.usgs.gov/resource/birds/ageduck/index.htm.

Kear, J. 2005. *Bird Families of the World: Ducks, Geese and Swans*. Oxford University Press, Oxford, UK.

Keller, V. 2014. From wintering to breeding: abundance and distribution of breeding Red crested Pochards *Netta rufina* in Switzerland. *Ornithol Beob* 111: 35–52.

Krakauer, A.H. & Kimball, R.T. 2009. Interspecific nest parasitism in galliform bird. *Ibis* 151: 373–381.

Lyon, B.E. & Eadie, J.McA. 1991. Mode of development and interspecific avian brood parasitism. *Behavioral Ecology* 2: 309–318.

Mednis, A. 1968. Biologija gnezdovanija utok na ozera Engures. *In* H.A. Mihelsons (ed.), *Ekologia Vodopalsvasjuscich ptic Latvii,* pp. 85–108. Institute of Biology, Riga, Latvia. [In Latvian.]

Mlíkovský, J. & Buřič, K. 1983. *Die Reiherente (Aythya fuligula)*. Die neue Brehm-Bücherei, Vol. 556. A. Ziemsen Verlag, Wittenberg Lutherstadt, Germany.

Musil, P. & Neužilová, Š. 2009. Long-term changes in duck inter-specific nest parasitism in South Bohemia, Czech Republic. *Wildfowl* (Special Issue No. 2): 176–183.

Neužilová, Š. & Musil, P. 2010. Inter-specific egg recognition among two diving ducks species, Common Pochard (*Aythya ferina*) and Tufted Duck (*Aythya fuligula*). *Acta Ornithologica* 45: 59–65.

Owen, M. & Black, J.M. 1990. *Waterfowl Ecology*. Blackie, Glasgow and London, UK.

Payne, R.B. 1977. The ecology of brood parasitism in birds. *Annual Review of Ecology and Systematics* 8: 1–28.

Rohwer, F.C. & Freeman, S. 1989. The distribution of conspecific nest parasitism in birds. *Canadian Journal of Zoology* 67: 239–253.

Sayler, R.D. 1992. Ecology and evolution of brood parasitism in waterfowl. *In* B.D.J. Batt, A.D. Afton, M.G. Anderson, C.D. Ankney, D.H. Johnson, J.A. Kadlec & G.L. Krapu (eds.), *Ecology and Management of Breeding Waterfowl*, pp. 290–322. University of Minnesota Press, Minneapolis, USA.

Smrček, M. 1981. *Hnízdní ekologie poláka chocholačky (Aythya fuligula), poláka velkého (Aythya ferina), kachny divoké (Anas platyrhynchos), kopřivky obecné (Anas strepera) a zrzohlávky rudozobé (Netta rufina) v jižních Čechách*. M.Sc. thesis, Charles University, Praha, Czech Republic. [In Czech.]

Sorenson, M.D. 1992. Comment: Why is conspecific nest parasitism more frequent in waterfowl than in other birds? *Canadian Journal of Zoology* 70: 1856–1858.

Sorenson, M.D. 1997. Effects of intra- and interspecific nest parasitism on a precocial host, the canvasback, *Aythya valisineria*. *Behavioral Ecology* 8: 153–161.

Sorenson, M.D. 1998. Patterns of Parasitic Egg laying and Typical Nesting in Redhead and Canvasback Ducks. *In* S.I. Rothstein & S.K. Robinson (eds.), *Parasitic Birds and their Hosts. Studies in Coevolution*, pp. 357–375. Oxford University Press, Oxford, UK.

Šťastný, K. & Hudec, K. (eds.) 2016. *Fauna ČR. Ptáci 1*. Academia, Praha, Czech Republic.

Weller, M.W. 1959. Parasitic egg laying in the Redhead *(Aythya americana)* and other North America Anatidae. *Ecological Monographs* 29: 333–365.

Yamauchi, A. 1995. Theory of evolution of nest parasitism in birds. *The American Naturalist* 145: 434–456.

Yom-Tov, Y. 1980. Intraspecific nest parasitism in birds. *Biological Reviews* 55: 93–108.

Photograph: Tufted Duck and Red-crested Pochard ducklings just hatched in a single clutch in South Bohemia, Czech Republic, by Petr Musil.

From field to museum tray: shrinking of the Mallard *Anas platyrhynchos*

MURRAY WILLIAMS

68 Wellington Road, Paekakariki 5034, New Zealand.
E-mail: murraywilliamsnz@outlook.com

Abstract

Four commonly-recorded body measurements were collected from fresh specimens of Mallard *Anas platyrhynchos* and of captive-raised Mallard × Grey Duck *A. superciliosa* hybrids, and the same specimens were measured again 25+ years later to determine the extent of tissue shrinkage. There was little shrinkage in bill length but average shrinkage in bill widths, tarsus lengths and wing lengths were between 3.0–4.7%, with changes in bill width and tarsus length being the most variable and extreme. Correction values to allow the combining of field and museum specimen measurements are provided.

Key words: Anatidae, *Anas platyrhynchos,* Mallard, measurements, specimen shrinkage.

Avian biometric data, whether from field or museum specimens, or both, have assisted studies seeking, for example, sex and age discrimination (*e.g.* Ó hUallacháin & Dunne 2010), population or taxonomic differentiation (*e.g.* Weidinger & van Franeker 1998; Robertson & Wareham 1994), evaluations of environmental responses (*e.g.* Gardner *et al.* 2009), and detection of latitudinal influences in avian body size (*e.g.* Engelmoer & Roselaar 1998; Graves 1991). However, detecting small differences in measurements can be corrupted by the shrinking of tissues as fresh specimens dry during the preservation process, especially of those body parts where the bone is overlain by flesh or connective tissue not removed when the specimen is prepared. Thus, the mixing of measurements from fresh or field specimens with those from dried museum specimens is problematic unless this shrinkage can be accounted for.

Several studies have identified the extent of tissue shrinkage that affects the commonly-taken measurements of bill, legs, wing and tail, for example Kinsky & Harper (1968) in prions *Pachyptila* sp., Harris (1980) and Ewins (1985) in Puffins *Fratercula arctica* and Black Guillemots *Cepphus grylle* respectively, Fjeldså (1980) in grebes *Podiceps* sp., Jenni & Winkler (1979) and Winker (1993) for passerines and Greenwood (1979), Green (1980) and Engelmoer *et al.* (1983) in small waders (Charadriiformes). These studies identified a considerable variability in the extent of shrinkage in individuals of the same species and highlighted the need for mean shrinkage estimates to be determined from substantial sample sizes. Collectively, they also suggest that the extent of shrinkage may differ between species within avian families but may be more consistent across species of similar size.

Wilson & McCracken (2008) provided correction values for measurements from specimen skins of a small (350–550 g) duck, Cinnamon Teal *Anas cyanoptera*, for comparing them with measurements obtained from live birds. This appears to be the only such assessment for Anseriformes despite some published compilations of waterfowl measurements, either freely combining field and museum measurements (*e.g.* some species accounts in Kear 2005) or, more rarely, listing both side by side (*e.g.* Marchant & Higgins 1990). I therefore assessed tissue shrinkage for larger (900–1,300 g) ducks by comparing fresh and dry lengths for wings of wild Mallard *Anas platyrhynchos*, and for bill and tarsus measurements taken from captive-raised Mallard × Grey Duck *Anas superciliosa* hybrids.

Methods

Source of specimens

Wings were obtained from wild Mallards shot by New Zealand hunters in May 1991. After collection, the sex of the bird was determined from the wing samples, according to whether the white bar anterior to the speculum extended onto the greater tertial coverts (females) or not (males; Carney 1992), but were not aged. The wings were measured by the author, muscle tissue between the ulna and radius was then removed and each wing attached to a cardboard backing sheet and air-dried. These wings were lodged in Museum of New Zealand Te Papa Tongarewa, Wellington as specimen numbers OR028752–OR028908 where 95 (48 females, 47 males) were re-measured 25 years later, again by the author.

Mallard × Grey Duck hybrids were bred in captivity 1967–1972 at Mt Bruce Native Bird Reserve near Masterton, New Zealand. Freshly-killed specimens were sexed by cloacal examination (Mosby 1963), weighed, measured (as below), prepared as specimen skins or pelts, and subsequently air-dried. A representative selection of F1, F2 and F3 hybrids was lodged in Museum of New Zealand Te Papa Tongarewa, Wellington as specimen numbers OR027923–31 and OR028494–568, where 60 specimens (30 females, 30 males) were re-measured by the author in 2016–17.

Mean weights (g) in May of wild Mallard in New Zealand are: males = 1,193 ± s.d. 104, n = 593; females = 1,075 ± s.d. 105, n = 336 (M. Williams, unpubl. data) and mean weights (g) in May of Grey Duck in New Zealand are: males = 1,054 ± s.d. 92, n = 234; females = 934 ± s.d. 84, n = 177 (Williams 2017). Mean weights (g) of the studied hybrids at death (mostly in May and August) were: males = 1,133 ± s.d. 141, n = 30; females = 1,041 ± s.d. 116, n = 30.

Measurement procedure

Wings of wild Mallard were measured while fresh (to nearest 1 mm) following Baldwin *et al.* (1931), as reported by Gurr (1947), using a steel ruler, and re-measurement was done the same way.

Measurements from freshly-killed Mallard × Grey Duck hybrids were of bill length (exposed culmen), bill width and tarsus length, all measured to 0.1 mm using a Vernier calliper. Bill and tarsus length measurements followed Baldwin *et al.* (1931), as reported by Gurr (1947), and bill width was measured at the gape, immediately below the culmen's proximal point of measurement. Re-measurement was conducted in the same way.

A paired *t*-test was used to compare differences between fresh and dry measurements. Pearson correlation values were used to examine the relationship between body mass and the percent shrinkage of bill and tarsus measurements.

Results

Significant reductions in wing and tarsus lengths, and also in bill widths, occurred after drying whereas bill lengths were little changed (Table 1).

Although the lengths of three dried wings were measured as being longer than when fresh (two by 1 mm; one by 2 mm), and a fourth wing was recorded as being unchanged, lengths of the remaining 91

Table 1. Mean (± s.d.) fresh and dry wing lengths (mm) of Mallards, and bill lengths, bill widths and tarsus lengths (mm) of Mallard × Grey Duck hybrids, with results of *t*-tests (analysing differences between the fresh and dry measurements), the percent shrinkage of the tissue, and correction values for converting dry measurements to fresh measurements. *** = $P < 0.0001$.

	n	Fresh (mm) mean ± s.d.	Dry (mm) mean ± s.d.	*t*	% Shrinkage mean ± s.d.	Correction factor ± 95% C.I.
Wing length						
Males	47	277.5 ± 11.2	269.2 ± 10.3	9.92***	3.0 ± 2.0	1.031 ± 0.006
Females	48	264.5 ± 9.6	256.0 ± 8.6	13.29***	3.2 ± 1.6	1.034 ± 0.006
Tarsus length						
Males	30	48.8 ± 1.6	47.3 ± 1.5	6.35***	3.0 ± 2.5	1.032 ± 0.010
Females	30	46.2 ± 1.9	44.5 ± 1.7	7.04***	3.6 ± 2.6	1.038 ± 0.010
Bill width						
Males	30	21.6 ± 0.9	20.6 ± 1.4	5.88***	4.7 ± 4.7	1.052 ± 0.018
Females	30	20.6 ± 0.8	19.8 ± 1.2	4.58***	3.8 ± 4.4	1.042 ± 0.017
Bill length						
Males	30	54.3 ± 2.7	54.6 ± 3.0	1.48	–0.5 ± 2.0	0.995 ± 0.007
Females	30	51.7 ± 2.5	51.9 ± 2.3	1.38	–0.5 ± 1.8	0.996 ± 0.006

dried wings were significantly shorter than their corresponding fresh measurement (by 1–19 mm, mean = 8.8 mm; t_{90} = 17.55, $P < 0.0001$).

The average percentage change in tarsus length following drying was similar to that of wing length, but more variable, as indicated by the larger s.d. (Table 1) and the range of the changes from an increase of 4.1% to a reduction of 9.9%. Five (8.3%) tarsus lengths increased after drying.

Changes in bill widths were the most extreme of all characters measured, and the most variable. In 11 (18%) of the specimens, shrinkage was > 8% (maximum 13.5%). Of 13 (22%) specimens having increased bill widths after drying, eight had extended by < 1% (overall maximum 3.7%).

The mean length of dried bills differed from that of fresh bills by 0.3 mm (0.5%), and had marginally increased rather than decreased in length. Whereas 42% of the dried bills measured were shorter than when fresh (mean = 0.7 mm, range = 0.1–2.0 mm) those longer were by an average 0.9 mm (range = 0.1–2.8 mm). Differences between the fresh and dry bill length measures however were not statistically significant (males: t_{29} = 1.48, P = 0.15, n.s.; females: t_{29} = 1.38, P = 0.18, n.s.; Table 1).

For none of the four characters measured was there a significant difference between males and females in the extent of tissue change on drying the specimens (wing length: t_{93} = 0.64, n.s.; tarsus length: t_{58} = 0.91, n.s.; bill width: t_{58} = 0.81, n.s.; bill length: t_{58} = 0.10, n.s.). Percentage reductions in tarsus and bill lengths, and in bill width, were not related to body mass (Pearson correlation values, tarsus length:

r_{58} = –0.193, n.s.; bill width: r_{58} = 0.226, n.s.; bill length: r_{58} = 0.027, n.s.).

Correction values (± 95% C.I.) for converting dry measurements to fresh measurements, and derived from combining male and female measurements for each character, are: wing length = 1.032 (± 0.004); tarsus length = 1.034 (± 0.007); bill width = 1.047 (± 0.050); and bill length = 0.995 (± 0.005).

Discussion

Reduction in wing, bill and tarsus lengths following specimen preparation and drying is now a well-recognised phenomenon (*e.g.* Table 2 for wing length measures), matched also with the realisation that repeat measurements by the same measurer, or measurements made by different measurers, are variable (*e.g.* Nisbet *et al.* 1970; Ewins 1985; Barrett *et al.* 1989). Both can produce confounded interpretations when attempting to discriminate small measurement differences. Herremans (1985) has suggested that, where tissue shrinkage or measurement variability is small, the range of normal variation within a population may be sufficient to buffer these effects and allow the combining of fresh and specimen measurements for most purposes. This perspective of convenience, nevertheless, ignores that real differences were detected.

In this study of large ducks, shrinkage significantly affected wing and tarsus lengths, and bill width at the gape, but not bill length. Compared to the shrinkage recorded for the smaller Cinnamon Teal (Wilson & McCracken 2008), mean wing length shrinkage was approximately 50%

Table 2. Examples of shrinkage in wing length on dried specimens of various species. Species weights (g) from Dunning (2007).

Species		Mean weight	Mean wing length (mm)	% Shrinkage mean ± s.d.	Source
Mallard	*Anas platyrhynchos*	1080	271	3.1 ± 1.8	This study
Cinnamon Teal	*Anas cyanoptera*	383 (♂) 372 (♀)	194 189	2.12 ± 2.24 2.17 ± 1.99	Wilson & McCracken 2008
Oystercatcher	*Haematopus ostralegus*	526	~260	2.7 ± 0.9	Engelmoer *et al.* 1983
Horned Grebe	*Podiceps auritalis*	453	140	1.02	Fjeldså 1980
Rook	*Corvus frugilegus*	450	308	1.24	Knox 1980
Puffin	*Fratercula arctica*	381	157	1.10	Harris 1980
Black Guillemot	*Cepphus grylle*	378	162	1.55	Ewins 1985
Dunlin	*Chalidris alpina*	56	114	1.8 ± 0.9	Engelmoer *et al.* 1983
Tennessee Warbler	*Vermivora peregrina*	8.9	62	1.44	Winker 1993

greater and more consistent across the specimens sampled. Tarsus shrinkage was similar however (3.4–3.9% in Cinnamon Teal; 3.6% in this study). Mean bill length (of exposed culmen) shrank by *c.* 1.1% in the teal, but was less than half of that for the larger ducks in this study and with slightly less variability.

Bill width (at gape) and tarsus length are two measurements that are problematic on dry specimens (*e.g.* Kinsky & Harper 1968; Fjeldså 1980). Distortion during the drying process of the flaccid bill flap and, especially, of the thick leg skin in the notch between femur and tarsus, make both measurements variable and poor indicators of the measurements from live specimens. Not surprisingly, both characters returned the most extreme and the most variable measurements from the dried specimens, as they did from the Cinnamon Teal (Wilson & McCracken 2008) where bill width measurement was taken below the nares.

It was pointed out in review that the correction values for converting dry measurements to fresh measurements are, themselves, only estimates, and that their application to small numbers of specimens and when seeking to correct for small measurement changes, may introduce additional bias. Confidence intervals, if provided with the correction values, may indicate the possible magnitude of any such bias. In addition, correction for small measurement changes can also be compromised by the generally undeclared variation in measurement consistency within or between measurers (Barrett *et al.* 1989).

The results of this study are from the largest waterfowl yet studied, and perhaps from the largest bird yet reported upon. Wing shrinkage recorded for 350–500 g birds (Table 2) appears to be unrelated to body size. However, more examples of specimen shrinkage from a diverse range of species may suggest otherwise. In the meantime, species-specific assessments are still required to allow combined use of measurements from live and dried museum specimens.

Acknowledgements

I acknowledge with gratitude the workspace and library access provided by the School of Biological Sciences, Victoria University of Wellington, New Zealand, and I thank Colin Miskelly, curator of birds at Museum of New Zealand Te Papa Tongarewa for access to specimen skins. I am also grateful to Jeremy Greenwood and an anonymous referee, for the perspectives offered and the discipline demanded.

References

Baldwin, S.P., Oberholser, H.C. & Worley, L.G. 1931. Measurements of birds. *Scientific Publications of the Cleveland Museum of Natural History* 2: 1–165.

Barrett, R.T, Peterz, M., Furness, R.W. & Durinck, J. 1989. The variability of biometric measurements, *Ringing & Migration* 10: 13–16.

Carney, S.M. 1992. *Species, Age and Sex Identification of Ducks using Wing Plumage*. US Department of Interior, US Fish & Wildlife Service, Washington D.C., USA.

Dunning, J.B. 2007. *Handbook of Avian Body Masses. Second Edition*. CRC Press, Florida, USA.

Engelmoer, M. & Roselaar, K. 1998. *Geographic Variation in Waders*. Springer, Dordrecht, The Netherlands.

Engelmoer, M., Roselaar, K., Boere, G.C. & Nieboer, E. 1983. Post-mortem changes in measurements of some waders. *Ringing & Migration* 4: 245–248.

Ewins, P.J. 1985. Variation in Black Guillemot wing lengths post-mortem and between measurers. *Ringing & Migration* 6: 115–117.

Fjeldså, J. 1980. Post-mortem changes in measurements of grebes. *Bulletin of the British Ornithologist's Union* 100: 151–154.

Gardner, J.L., Heinsohn, R. & Joseph, L. 2009. Shifting latitudinal clines in avian body size correlate with global warming in Australian passerines. *Proceedings of the Royal Society, B.* doi:10.1098/rspb.2009.1011.

Graves, G.R. 1991. Bergmann's rule near the equator: Latitudinal clines in body size of an Andean passerine bird. *Proceedings of National Academy of Science USA* 88: 2322–2325.

Green, G.H. 1980. Decrease in wing length of skins of Ringed Plover and Dunlin. *Ringing & Migration* 3: 27–28.

Greenwood, J.G. 1979. Post-mortem shrinkage of Dunlin *Calidris alpina* skins. *Bulletin of the British Ornithologists Club* 99: 143–145.

Gurr, L. 1947. Measurements of birds. *New Zealand Bird Notes* 2: 57–61.

Harris, M.P. 1980. Post-mortem shrinkage of wing and bill of Puffins. *Ringing & Migration* 3: 60–61.

Herremans, M. 1985. Post-mortem changes in morphology and its relevance to biometrical studies. *Bulletin of the British Ornithologist's Club 105*: 89–91.

Jenni, L. & Winkler, R. 1989. The feather length of small passerines: a measurement for wing length in live birds and museum skins. *Bird Study* 36: 1–15.

Kear, J. (ed.). 2005. *Ducks, Geese and Swans of the World.* Oxford University Press, Oxford, UK.

Kinsky, F.C. & Harper P.C. 1968. Shrinkage of bill width in skins of some *Pachyptila* species. *Ibis* 110: 100–102.

Knox, A. 1980. Post-mortem changes in wing lengths and wing formulae. *Ringing & Migration* 3: 29–31.

Marchant, S. & Higgins, P. (eds). 1990. *Handbook of Australian, New Zealand and Antarctic Birds. Volume 1, Part B.* Oxford University Press, Melbourne, Australia.

Mosby, H.S. (ed.). 1963. *Wildlife Investigational Techniques. Second edition.* The Wildlife Society, Washington, USA.

Nisbet, I.C.T., Baird, J., Howard, D.V. & Anderson, K.S. 1970. Statistical comparison of wing lengths measured by four observers. *Bird Banding* 41: 307–308.

Ó hUallacháin, D. & Dunne, J. 2010. Analysis of biometric data to determine the sex of Woodpigeons *Columba palumbus. Ringing & Migration* 25: 29–32.

Robertson, C.J.R. & Wareham, J. 1994. Measurements of *Diomedia exulans antipodensis* and *D. e. gibsoni. Bulletin of the British Ornithologist's Club* 114: 132–134.

Weidinger, K. & van Franeker, J.A. 1998. Applicability of external measurements to sexing of the Cape Petrel *Daption capense* at within-pair, within-population and between-population scales. *Journal of Zoology* 245: 473–482.

Williams, M. 2017. Weights and measurements of *Anas superciliosa* in New Zealand. *Notornis* 64: 162–170.

Wilson, R.E. & McCracken, K.G. 2008. Specimen shrinkage in Cinnamon Teal. *The Wilson Journal of Ornithology* 120: 390–392.

Winker, K. 1993. Specimen shrinkage in Tennessee Warblers and "Traill's" Flycatchers. *Journal of Field Ornithology* 64: 331–336.

Differential flight responses of spring staging Teal *Anas crecca* and Wigeon *A. penelope* to human *versus* natural disturbance

THOMAS BREGNBALLE*, CHARLOTTE SPEICH, ANDERS HORSTEN & ANTHONY D. FOX

Department of Bioscience, Aarhus University, Kalø, Grenåvej 14, DK-8410 Rønde, Denmark.
*Correspondence author. E-mail: tb@bios.au.dk

Abstract

Observations made of disturbance to spring staging Wigeon *Anas penelope* and Teal *A. crecca* by human and "natural" (non-human) stimuli at a restored wetland in the Skjern River delta, Denmark, were analysed to inform future management of human access to the site. The effects of human activity (anglers, cyclists, farming activity) on the flight responses and displacement distances of ducks within uniform habitat along a public path were compared with the birds' reaction to natural stimuli such as mammals or birds of prey. Excluding the controlled disturbance by a pedestrian, undertaken as part of the study, the main cause of flushing in Wigeon was a response to the movements of birds of prey and other birds, especially Lapwings *Vanellus vanellus* performing flight displays. For Teal, birds of prey accounted for around half of the flushes, with other birds accounting for one third of the flushes. Wigeon and Teal were displaced significantly farther by human activities than by natural causes. We tested whether the ducks reacted differently to natural disturbances shortly after disturbance by a pedestrian by comparing response patterns to natural stimuli within the first hour following disturbance from a passing pedestrian with their response patterns in the absence of pedestrians, but found no evidence to suggest that they did so. In our study area, Wigeon used land for feeding and water as predator-escape habitat; 23% of the 144 observed take-offs of Wigeon were from water but 68% of the landings were on water. Of the 83 observations of flushed Teal, 56% flushed from water and 51% landed on water.

Key words: displacement distance, disturbance, human disturbance, public footpath, waterbirds, waterfowl refuges, wetland.

Despite domestic and international legislation to protect wetland habitats, rates of loss and degradation of wetlands have been rapid during the 20th and early 21st centuries (Davidson 2014). Even well-protected wetlands are facing pressures from multiple competing uses, especially from increasing urban human populations that wish recreational access to such areas. Many features of modern recreational use and nature conservation interests are directly incompatible with each other. For instance, many recreational activities cause waterbirds to suffer increased vigilance, loss of feeding, enhanced energy expenditure, or a combination of these to the extent that they desert a site or suffer reductions in reproductive success or survival (Madsen & Fox 1995, 1997; Livezey *et al.* 2016). On the other hand, it is essential that there is public access to protected wetlands to ensure societal support for their protection. International legislation also enshrines the sustainable use of such areas and includes non-destructive recreational use of wetlands as part of their guiding principles (*e.g.* in the Ramsar Convention; UNESCO 1994). One important mechanism for reducing conflict is to zone recreational activities on protected wetlands, thus providing birds with refuge areas from loci of major disturbance, but this requires an understanding of the distances at which waterbirds react to sources of disturbance and the distances over which they are displaced (*e.g.* Fox & Madsen 1997; Livezey *et al.* 2016).

Such an understanding is especially important with regard to restored wetlands, where large investments of public and/or private money to recreate habitats heighten the expectations of local communities with regard to the potential recreation and other demands placed upon a new and novel resource. Fostering public support for wetland restoration is essential for the long-term sustainable management and use of

such restored wetlands because of the loss of traditional activities (such as waterbird hunting or fishing) prior to reinstatement (Scholte *et al.* 2016). Many case studies have demonstrated that a lack of effective coordination between responsible managing organisations often leads to direct competition that favours cultural ecosystem services (especially public access and tourism) to the detriment of habitat services (*i.e.* biodiversity conservation; Cohen-Shacham *et al.* 2015). Because of such pressures to integrate human access requirements with nature conservation management objectives on restored wetlands, it is vital that we have a good understanding of how individual waterbird species respond to pedestrian human access, and precisely how these may interact with natural sources of disturbance to affect the attractiveness to wildlife of a reinstated wetland.

Here we present the results of a study on the causes of displacement (*i.e.* flushes) of Wigeon and Teal, including an assessment of how these two species respond to human activities within or on the periphery of the wetland. The study was carried out in the lower Skjern River in Denmark, which consists of 22 km² of restored wetlands including lakes, shallow wetlands and flooded wet grassland (Petersen *et al.* 2007; Bregnballe *et al.* 2014). Our objectives were to describe responses of spring-staging dabbling ducks to natural disturbance stimuli before and after the passage of a single person walking on a path along habitat used by these birds. Specifically, we wished to test the hypothesis that Wigeon *Anas penelope* and Teal *A. crecca* (the commonest spring-staging wetland species at the site) responded by flying greater distances to settle after being disturbed by human activities than in response to natural sources of disturbance. Secondly, we sought to find whether these two species responded differently to disturbance soon after having been disturbed by a passing pedestrian, compared to their responses in the absence of pedestrians (see Smit & Visser 1993), for informing policy on future human access to such areas of a restored wetland complex.

Study area and methods

The study area was a section of continuous wet grassland habitat within a much larger wetland restoration complex along the lower River Skjern, west Jutland, Denmark (55°55'N, 8°25'E; Bregnballe *et al.* 2009, 2014). The human activities that in some cases caused disturbance to Wigeon and Teal included anglers walking along the river, cyclists (biking outside the footpath) and noise originating from agricultural activities on adjacent farmland. The behavioural response of dabbling ducks to the controlled disturbance from a pedestrian using the public footpath in the study area was described in Bregnballe *et al.* (2009). The study site could be overlooked from a fixed observation point on a dike (without disturbing the birds) and most dabbling ducks present could see pedestrians on the footpath which runs 450 m along the study site (see map in Bregnballe *et al.* 2009). When undisturbed, dabbling ducks feed throughout most parts of the study area, so we assumed that the entire area offered suitable habitat. A very detailed map of the

studied part of the wetland was drawn from aerial photographs, with visible water edges and vegetation features used to divide the study site into a large number of sub-areas. Displacement distances were measured as the distance from the centre of the birds' take-off sub-area to the centre of their landing sub-area. Each displacement distance recorded was then grouped into one of six intervals: 0–50 m, 50–100 m, 100–150 m, 150–200 m, 200–300 m and > 300 m. We observed daytime reactions of Wigeon and Teal to human activity and to the presence of other birds and mammals, during 25 mornings in March and April of 2003 and 2004. The path along the study site was closed to public access during these times to reduce uncontrolled disturbance of the ducks by pedestrians. The public had free access to the path along the study site throughout the day on days when no experiments were undertaken, as well as during the rest of the day after the observations associated with the controlled disturbance had ended. One person acted as the pedestrian whilst a second person, concealed from the birds' view, made behavioural observations from the dike. The behavioural observations were made using binoculars and a telescope and dictated onto a recorder. After the observer had mapped the waterbirds present at the study site, the pedestrian approached the study site, walking at normal speed interspersed by short stops lasting up to approximately 3 min, simulating someone observing wildlife. As soon as the pedestrian was at least 100 m from the study area, they were rendered invisible to the birds because of tall vegetation.

Causes of Wigeon or Teal taking flight (hereafter called "flushing") were grouped into six categories: human (anglers, cyclists elsewhere than on the footpath, farmers), birds of prey, Lapwing *Vanellus vanellus*, other birds (larger than *c.* 100 g, so omitting the smaller passerines), mammals and "unknown". Birds of prey which the ducks always or sometimes flushed from included Osprey *Pandion haliaetus*, Marsh Harrier *Circus aeruginosus*, Hen Harrier *C. cyanus*, Buzzard *Buteo buteo*, Sparrowhawk *Accipiter nisus*, Goshawk *A. gentilis*, Kestrel *Falco tinnunculus*, Merlin *F. columbarius* and Short-eared Owl *Otus flammeus*. Other birds included Cormorant *Phalacrocorax carbo*, Grey Heron *Ardea cinerea*, Spoonbill *Platalea leucorodia*, geese *Anser* sp., Curlew *Numenius arquata* and two gull *Larus* species (Herring Gull *Larus argentatus* and Great Black-backed Gull *L. marinus*). Mammals were Red Fox *Vulpes vulpes* and Roe Deer *Capreolus capreolus* (ducks did flush from deer on occasion). We compared the frequency distribution of the causes of flushing in the non-disturbed situation with the frequency distribution of causes within 15–70 min after the manipulated disturbance by the pedestrian had ended (*i.e.* the human was no longer visible to the birds), using two-sample Z tests to test for differences between these two sets of data for each response category. Only one controlled disturbance event was carried out per day.

Wilcoxon two-sample tests were used to compare displacement distances for flushing caused by human disturbance with those caused by natural disturbances, across each of the six distance categories described above. The natural disturbance category consisted of flushing caused by "birds of prey", "other birds", "mammals" or "Lapwing".

Results

In the flight-response test, Wigeon often flushed without any evident cause, both before and after the controlled human disturbance (26% and 32% of all flushing events, respectively, $n = 93$; Fig. 1a). Lapwing, other birds and birds of prey each contributed 15–26% of flushes, whereas humans (present further away than the public footpath) and mammals together accounted for 8% of the flushes by Wigeon both before and after the controlled disturbance from the footpath. For Teal, birds of prey accounted for 54% of the flushes ($n = 93$) before the controlled disturbance but only 25% after the disturbance (Fig. 1b), whereas other birds accounted for 29% of the flushes both before and after disturbance. The number of flushes attributable to Lapwings more than doubled from 8% to 21% before and after the disturbance event. There were no significant differences (at $P < 0.05$) between the proportions of flushes in each category before and after manipulated human disturbances for Wigeon, but Teal reacted significantly more frequently before the controlled disturbance from the pedestrian than afterwards ($Z = 1.98$, $P = 0.027$; Fig. 1).

Displacement distances for Wigeon and Teal were significantly longer when flushes were induced by human disturbance than when flushes were the result of natural causes (Fig. 2). In the 74% of cases where Wigeon were displaced by natural causes they landed within 50 m of their original

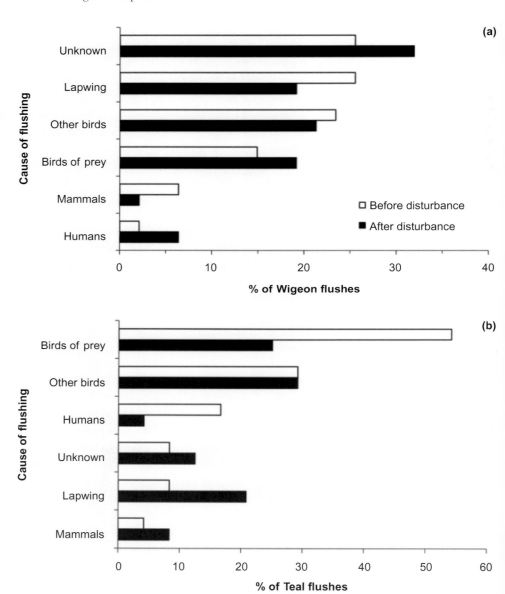

Figure 1. Cause of flushes for (a) Wigeon (n = 93 incidents) and (b) Teal (n = 93) observed at Skjern River, western Jutland, Denmark both prior to and following controlled disturbance by a human walking along a footpath adjacent to the feeding area. Differences in the response of Teal to "birds of prey" before and after the human disturbance were statistically significant (Wilcoxon two-sample test: Z = 1.98, P = 0.027).

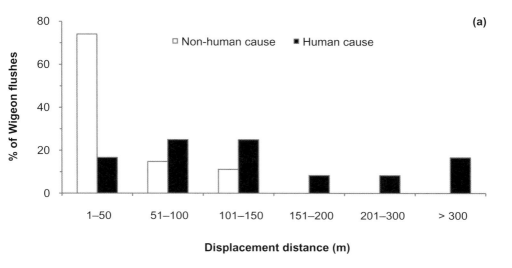

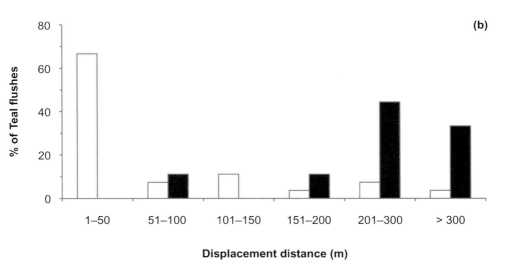

Figure 2. Frequency distributions of displacement distances following flushes induced by human disturbance and from natural causes for (a) Wigeon, and (b) Teal at the Skjern River study site in western Jutland, Denmark. For Wigeon sample sizes were 12 records of displacement distance for flushes caused by humans and 27 records for flushes caused by natural causes. For Teal sample sizes were 9 and 27, respectively. Displacement distances were significantly greater for human disturbance than for "natural" disturbance in both Wigeon (Wilcoxon two-sample test: $U_s = 189.5$, $P < 0.01$) and Teal ($U_s = 199$, $P < 0.01$).

location, whereas they did so on only 17% of the occasions when flushed by human disturbance (Fig. 2a; see figure caption for sample sizes). Wigeon flushed by human disturbance flew > 150 m in 33% of all cases, but flights of > 150 m were never recorded when Wigeon were flushed by natural causes. Similar flight responses were apparent for Teal (Fig. 2b), where 67% of flushes induced by natural causes were short (< 50 m), and flights were always longer than this in response to human disturbance. Long flights (> 200 m) were recorded in > 78% of the cases when flushes of Teal were caused by human disturbance compared to only 11% of the cases when flushes were induced by natural causes. There were also indications that "birds of prey" elicited shorter displacement distances at our study site than human disturbance despite harriers attacking and pursuing Teal. For instance, for Teal 75% of 16 displacement distances caused by raptors were of < 100 m, whereas only 11% of nine displacement distances caused by humans were < 100 m. For Wigeon the proportions were 70% (*n* = 12) and 42% (*n* = 10), respectively.

Amongst all 144 observations of Wigeon flushed from all causes combined, 23% were from water, since most active feeding by this species occurred on terrestrial wet grassland. However, 68% of the time they landed on water after being flushed. Of the 83 observations of flushed Teal, 56% flushed from water and 51% landed on water.

Discussion

These results indicated that there was no difference in the response of Wigeon to different sources of disturbance before and after disturbance by a pedestrian, and that Teal were more likely to react prior to such disturbance, which suggests there was no elevated response to particular sources of disturbance after birds were exposed to manipulated disturbance from a pedestrian. The results showed significantly longer flushing distances before resettling after human disturbance (excluding the manipulated disturbance caused by the pedestrian) compared to those generated by all natural sources of disturbance, suggesting both an enhanced energetic cost and possibly also a greater displacement from favoured feeding areas when disturbed by humans compared to other stimuli.

The majority of Teal flushes were responses to birds of prey, other birds and "human activity" (*e.g.* anglers, cyclists). Marsh Harrier and Hen Harrier induced most predator flushes; we observed several attacks on Teal by harriers and both species of harriers attack ducks, despite their established preference for smaller prey (Génsbøl 2004). More specialised avian duck predators such as Peregrine *Falco peregrinus* and White-tailed Eagle *Haliaetus albicilla* were not observed during our observation period, but both species were present in the area. The main stimulus in the "other birds" category was the Great Black-backed Gull, a species observed attacking ducks along Skjern River (J.P. Hounisen, pers. comm.) and a probable explanation for why "other birds" induced flushing amongst Teal. Unlike Wigeon, "Lapwing" accounted for a small proportion of flushes in Teal. This difference may reflect the tendency for Wigeon to feed on land to a greater extent than Teal (Bregnballe *et al.* 2009), where they may be more sensitive to disturbance and perhaps have a lower stimulus threshold, making them more apt to take off in response to stimuli that we failed to detect (Mayhew & Houston 1987). Wigeon often flushed in response to "Lapwing" and many "unknown cause" flushes probably were associated with "Lapwings" where we failed to detect the true stimulus. There may be several explanations for this, including the fact that Lapwing silhouettes are somewhat similar to that of a broad-winged bird of prey and that displaying Lapwings are noisy in spring when their erratic display flights (with many rapid turns and dives) present sudden and unpredictable movements. It has been suggested that birds have evolved anti-predator responses to generalised, threatening stimuli such as loud noises and rapidly approaching objects (Frid & Dill 2002). Secondly, foraging Wigeon were often seen near Lapwings, since both species select for shorter grass swards, whereas Teal remained in shallow waters where they rarely encountered displaying Lapwings. Thirdly, Wigeon occurred in large flocks, whereas Teal often were observed foraging in small groups and flushing probability and flushing distance from disturbance has been shown to be positively associated with increasing flock size (review in Smit & Visser 1993; Bregnballe *et al.* 2009).

Waterbirds are subject to multiple sources of disruption to their daily activities, so responses to a specific locus of disturbance is likely to be a function of the presence and activity of other sources of disturbance and their frequency. The relative importance of any one source of disturbance is likely to be highly site-specific because of the array of disruptive activities to which birds are exposed at a given site. Hence, these findings may not apply to other locations, other times of year, or even to other times of the day. We also conclude that Teal flushes were, to a large extent in this study, restricted to situations where the source of the disturbance posed a threat to the birds, whereas Wigeon seemed to take off more frequently when the source of disturbance did not necessarily appear to pose a threat to the individuals concerned.

In this study, Wigeon and Teal flew significantly farther following human disturbances than after natural disturbances, but sample sizes were small. This was similar to the results presented by Béchet *et al.* (2004), who found that distances flown by spring staging Greater Snow Geese *Anser caerulescens atlantica* were longer after scaring and hunting than after natural forms of disturbances (including raptors and other predators). In our study area most types of human activities that caused disturbance could be characterised as occasional, unpredictable events, such as a sudden loud noise from farming activity, a cyclist appearing where cyclist would not normally appear, or an angler suddenly becoming visible. It therefore seems likely that, given the likely high turnover of individuals during spring staging, such groups of dabbling ducks experience difficulties in rapidly adapting to such unpredictable potential disturbance stimuli in situations where they do not remain long enough to be able to predict the likely mortality risks associated with them.

The energy expenditure associated with escape flights increases with the distance flown and time spent flying (*e.g.* Birt-Friesen *et al.* 1989; Nolet *et al.* 2016). Waterbirds consume around 10–12 times the energy at basal metabolic rate when flying (*e.g.* Mooij 1992 for flying geese), which represents a four-fold increase over the energetic costs of foraging and a six-fold increase compared to roosting. Furthermore, the cost of sudden locomotion requires the rapid attainment of escape velocity and manoeuvrable flight to avoid capture (Blumstein 2003). In addition to the energetic costs of escape, flushed birds lose foraging time and net energy gain during displacement from optimal foraging areas.

Overall, despite small sample sizes, our results suggest that flushes caused by human disturbance are associated with a greater energetic cost than flushes from natural causes due to flight costs. Furthermore, costs of escape may be even higher for Wigeon, since our results showed they tended to land in habitat unsuitable for foraging (water) and consequently have to return to their foraging habitat before resuming feeding. For instance, human disturbance displaced Wigeon and Teal over longer distances before they resettled than was the case following flushes attributable to natural stimuli, and the choice of habitat to which they were displaced (*i.e.* open water) also suggested a greater loss of feeding time, in addition to the greater energetic costs of such a disturbance event when caused by humans. These results clearly show that human activities can have a disproportionate energetic cost for staging waterbirds, in this case during the prelude

to breeding. This suggests that if the primary site management objectives are to protect staging waterbirds (for example over strictly amenity interests), then human access should be managed in a sympathetic manner at the site (*e.g.* through the screening of public footpaths and creation of viewing hides) to achieve public access to wetlands and waterbirds without disruption to the birds' normal activities.

On a broader level, it is difficult to judge how likely the energetic costs of human displacement are to extend to influencing the survival and breeding success of the individuals affected. Other studies of waterbirds suggest that these birds may respond more rigorously to disturbance when the foraging costs are lower (*e.g.* Yasué 2006), which may suggest that behavioural responses may not directly reflect the potential fitness costs of human disturbance (see Gill *et al.* 2001 for a broader discussion). Nevertheless, there is abundant evidence that persistent human disturbance to dabbling duck and other waterbirds leads to permanent local displacement that represents a net loss to the "carrying capacity" of this part of the site (*e.g.* Madsen 1998). It is therefore important to establish clear conservation management planning priorities when designing future wetland restoration schemes to recognise the relative importance of protecting such staging waterbird populations in relation to other management priorities for the site after reinstatement. More results from studies of other species than those considered here would further support such development of management priorities and provide potential methods to avoid conflicts between management objectives.

Acknowledgements

We thank the National Forest and Nature Agency for allowing access to carry out this study on the wetland and Kent Livezey and an anonymous referee, for their constructive comments, which helped to improve on an earlier version of the manuscript.

References

Béchet, A., Giroux, J.-F. & Gauthier, G. 2004. The effect of disturbance on behaviour, habitat use and energy of spring staging snow geese. *Journal of Applied Ecology* 41: 689–700.

Birt-Friesen, V.L., Montevecchi, W.A., Cairns, D.K. & Macko, S.A. 1989. Activity-specific metabolic rate of free-living Northern Gannets and other seabirds. *Ecology* 70: 357–367.

Blumstein, D.T. 2003. Flight initiation distance in birds is dependent on intruder starting distance. *Journal of Wildlife Management* 67: 852–857.

Bregnballe, T., Speich, C., Horsten, A. & Fox, A.D. 2009. An experimental study of numerical and behavioural responses of spring staging dabbling ducks to human pedestrian disturbance. *Wildfowl* (Special Issue No. 2): 131–142.

Bregnballe, T., Amstrup, O., Holm, T.E., Clausen, P. & Fox, A.D. 2014. Skjern River Valley, Northern Europe's most expensive wetland restoration project: benefits to breeding waterbirds. *Ornis Fennica* 91: 231–243.

Cohen-Shacham, E., Dayan, T., de Groot, R., Beltrame, C., Guillet, F. & Feitelson, E. 2015. Using the ecosystem services concept to analyse stakeholder involvement in wetland management. *Wetlands Ecology and Management* 23: 241–256.

Davidson, N.C. 2014. How much wetland has the world lost? Long-term and recent trends in global wetland area. *Marine and Freshwater Research* 65: 934–941.

Fox, A.D. & Madsen, J. 1997. Behavioural and distributional effects of hunting disturbance on waterbirds in Europe: implications for refuge design. *Journal of Applied Ecology* 34: 1–13.

Frid, A. & Dill, L. 2002. Human-caused disturbance stimuli as a form of predation risk. *Conservation Ecology* 6: 11.

Génsbøl, B. 2004. *Rovfuglene i Europa*. Gyldendal, Copenhagen, Denmark. [In Danish.]

Gill, J.A., Norriss, K. & Sutherland, W.J. 2001. Why behavioural responses may not reflect the population consequences of human disturbance. *Biological Conservation* 97: 265–268.

Livezey, K.B., Fernández-Juricic, E. & Blumstein, D.T. 2016. Database of bird flight initiation distances to assist in estimating effects from human disturbance and delineating buffer areas. *Journal of Fish and Wildlife Management* 7: 181–191.

Madsen, J. 1998. Experimental refuges for migratory waterfowl in Danish wetlands. II. Tests of hunting disturbance effects. *Journal of Applied Ecology* 35: 398–417.

Madsen, J. & Fox, A.D. 1995. Impacts of hunting disturbance on waterbirds: a review. *Wildlife Biology* 1: 193–207.

Madsen, J. & Fox, A.D. 1997. The impact of hunting disturbance on waterbird populations – the concept of flyway networks of disturbance-free areas. *Gibier Fauna Sauvage* 14: 201–209.

Mayhew, P. & Houston, D. 1987. Feeding site selection by Wigeon *Anas penelope* in relation to water. *Ibis* 131: 1–8.

Mooij, J.H. 1992. Behaviour and energy budget of wintering geese in the Lower Rhine area of North Rhine-Westphalia, Germany. *Wildfowl* 43: 121–138.

Nolet, B.A., Kolzsch, A., Elderenbosch M. & van Noordwijk, A.J. 2016. Scaring waterfowl as a management tool: how much more do geese forage after disturbance? *Journal of Applied Ecology* 53: 1413–1421.

Scholte, S.S.K., Todorova, M., van Teeffelen, A.J.A. & Verburg, P.H. 2016. Public support for wetland restoration: what is the link with ecosystem service values? *Wetlands* 36: 467–481.

Smit, C.J. & Visser, G.J.M. 1993. Effects of disturbance on shorebirds: a summary of existing knowledge from the Dutch Wadden Sea and Delta area. *Wader Study Group Bulletin* 68: 6–19.

UNESCO 1994. Convention on Wetlands of International Importance especially as Waterfowl Habitat. United Nations Educational, Scientific and Cultural Organization (UNESCO), Office of International Standards and Legal Affairs, Paris, France. Available at http://www.ramsar.org/sites/default/files/documents/library/current_convention_text_e.pdf (last accessed 28 September 2017).

Yasué, M. 2006. Environmental factors and spatial scale influence shorebirds' responses to human disturbance. *Biological Conservation* 128: 47–54.

Photograph: Wigeon flushing, by Niels J.H. Andersen.

Wildfowl

Instructions for Authors

This information can also be found at http://wildfowl.wwt.org.uk/, together with PDF files of papers from earlier editions of *Wildfowl*.

Editorial policy

Wildfowl is an international scientific journal published annually by the Wildfowl & Wetlands Trust (WWT). It disseminates original material on the ecology, biology and conservation of wildfowl (*Anseriformes*) and ecologically-associated birds (such as waders, rails and flamingos), and on their wetland habitats. Research and review articles related to policy development and application are welcome. Material on habitat management is also sought, particularly where this is directed to the conservation of wildfowl and other wetland birds.

In all cases material should not have been published elsewhere or be subject to current consideration for publication in other journals.

Policy on ethics for ornithological research

Research projects submitted as papers for publication in *Wildfowl* must have proper regard for animal welfare and habitat conservation, and employ humane practices. Attention is drawn to the guidelines published in *Animal Behaviour* 61: 271–275 and on the journal's website: www.elsevier.com/wps/find/journaldescription. cws_home/622782/authorinstructions#6002. The impact of a particular study should be evaluated in terms of the possible gains in knowledge (and the practical use of this knowledge), weighed against potential adverse consequences for individuals, habitats or populations. Papers are considered for publication in *Wildfowl* solely on the condition that the work reported was undertaken within relevant legal statutes, or where work is carried out in areas lacking legislation or regulation, the work should conform to ethical standards expected in the UK.

The Editor reserves the right to review and reject papers on this basis.

Manuscripts

There is no page charge for published papers. All papers accepted for publication become the copyright of WWT. Colour figures can be included, but the authors or their organisations may be expected to cover the additional cost of the colour pages.

Initial submission

Manuscripts should be prepared in accordance with the Instructions to Authors to *Wildfowl*. The Editor reserves the right to modify manuscripts that do not conform to scientific, technical, stylistic or grammatical standards and minor alterations of this nature will normally be seen by authors only at the proof stage.

An electronic version of the paper, in MS Word, should be submitted as an email attachment to The Editor at wildfowl@wwt.org.uk. Submissions should be received by 15 May for publication in November of the same year. However, WWT reserves the right to postpone publication until the issue of the following year.

Types of paper

Standard papers

A standard paper should present the results of original research. The data must not have been published elsewhere, and the text should not normally be longer than 8,500 words. The format required for standard papers is described below.

Review papers

Papers on topical subjects of relevance to the journal's remit. Reviews are often designed to summarise a particular subject area and/or to stimulate debate and further research. They should be presented in a style similar to that of standard papers as far as possible, and should not exceed 8,500 words.

Short communications

Short communications present new information that is often obtained from preliminary research or as a by-product of larger projects. A flexible structure is opted for in the presentation of these papers, with the author including as much introductory, methodological and discussion material as is necessary to show the context and relevance of the communication. However, short communications should follow the standard format, be concise and limited to a maximum of 3,500 words.

Field surveys

Survey and expeditionary material that provides new data on particular species or wetlands of special interest for waterbirds is also acceptable. Field survey reports should be of general interest to the readers of *Wildfowl*, and are normally limited to a maximum of five printed pages. They should include reference to the location of more detailed information on the work undertaken. Survey and expedition reports must be written in *Wildfowl* style and follow the format of standard papers.

Manuscript format

Manuscripts should be typed in double line spacing with a generous margin (*c.* 3.25 cm) each side. Pages should be numbered consecutively, including those containing acknowledgements, references, tables and figure legends. It is preferable that authors prepare their main text in Microsoft Word (Garamond or Arial font) and graphs in Microsoft Excel. Manuscripts must be in English and spelling should conform to the *Concise Oxford Dictionary of Current English*. The passive voice is preferred; the active voice may be used only occasionally, typically to emphasise a personal opinion in the Introduction or Discussion.

Title page – this should contain:
A concise and informative title (as short as possible). Do not include the authorities for taxonomic names in the title.

A list of authors' names (commencing with the correspondence author), along with their contact address details that should be valid for the coming year. Use first names and subsequent initials, not just initials, for authors. For instance, Adrian D. Smith, not A. D. Smith. An E-mail address should be included for the first or corresponding author.

A running header of not more than 45 characters.

Abstract – this should state the main purpose of the paper and give the key results, conclusions and recommendations. The Abstract should not include references or speculation, and should not exceed 340 words. Please note that all text should be written in the "third person"; *i.e.* "A study was carried out ..." and not "I carried out a study of ...".

Key words – a list, in alphabetical order, of five words or short phrases, excluding words used in the title.

Introduction – this should give the background to the study, including the hypotheses being tested and the reasons why the study is thought to be worthwhile.

Methods – a concise description of data collection, analytical methods, and equipment used (where appropriate), in sufficient detail for the work to be repeated.

Results – the results of the analyses, drawing attention in the text to information provided in the tables and figures. Where appropriate, the Results should follow the order of fieldwork/ analysis presented in the Methods section.

Discussion – highlight the significance of the results in relation to the objectives for the work, and place them in the context of the broader scientific field. Where appropriate, resulting recommendations, *e.g.* for future management or research, should be clearly set out.

Acknowledgements – should be concise and appropriate.

References – when mentioned in the text, references should be listed in chronological order, separated by a semi-colon. Citation of work by one or two authors should be in full (Owen 1980; Bannister & Walker 1998), but where there are more than two authors, the citation should be abbreviated to *et al.* (Worden *et al.*) in the text. When different groups of authors with the same first author and date occur, they should be cited as (Thomson *et al.* 1991a, b).

In the reference list, the references should be given in alphabetical order by name of the first author, then by name of the second author (where only one co-author) and then by publication year. All authors' names should be quoted, with the journal name in full and in italics. For example:

Baldassarre, G.A. & Bolen, E.G. 1994. *Waterfowl Ecology and Management.* Wiley & Sons, New York, USA.

Béchet, A., Giroux, J.-F., Gauthier, G., Nichols, J.D. & Hines, J.E. 2003. Spring hunting changes the regional movements of migratory Greater Snow Geese. *Journal of Applied Ecology* 40: 553–564.

Follestad, A. 1994. Background for a management plan for geese in Norway. *NINA Utredning* 65: 1–78. [In Norwegian with English summary.]

Frederiksen, M., Fox, A.D., Madsen, J. & Colhoun, K. 2001. Estimating the total number of birds using a staging site. *Journal of Wildlife Management* 65: 282–289.

Nudds, T.D. 1992. Patterns in breeding waterfowl communities. *In* B.D.J. Batt, A.D. Afton, M.G. Anderson, C.D. Ankney, D.H. Johnson, J.A. Kadlec & G.L. Krapu (eds.), *Ecology and Management of Breeding Waterfowl*, pp. 540–567. University of Minnesota Press, Minneapolis, USA.

Owen, M. 1980. *Wild Geese of the World.* B.T. Batsford, London, UK.

Wetlands International. 2002. *Waterbird Population Estimates, 3rd edition.* Global Series No. 12. Wetlands International, Wageningen, The Netherlands.

Worden, J., Mitchell, C.R., Merne, O.J. & Cranswick, P.A. 2004. Greenland Barnacle Geese *Branta leucopsis* in Britain and Ireland: results of the international census, spring 2003. Unpublished report to JNCC, The Wildfowl & Wetlands Trust, Slimbridge, UK.

When referring to articles written in a language other than English, give the title in the original language. If the article is in a language other than English but contains an English summary, this should be stated in parentheses at the end of the citation, as shown for the Follestad (1994) paper above. References should be cited as "in press" only if the paper has been accepted for publication. Work not yet submitted for publication may be cited in the text and attributed to its author as "full author name, unpublished data".

Tables
Each table should be on a separate page, numbered and titled. Table headings should be concise and tables should be numbered consecutively in the text as Table 1, *etc.* Data should not be duplicated in both figure and table form. Footnotes should be listed as letters not numbers.

Horizontal rules should be used in the tables themselves; use vertical rules only when absolutely necessary. The horizontal lines should be kept to a minimum, ideally limited to single horizontal lines before and after column headings and at end of table (*e.g.* Simple 1 format in Microsoft Word), with blank rows used to separate information presented within a table where needed to help interpretation. Keep tables in portrait format if possible, to fit one column (width: 6 cm) or 1 page (width: 13 cm) of *Wildfowl*.

Figures
Each figure should be on a separate page, with figure headings listed on a separate sheet. Figures should be about 50% larger than final printed size. Figures should fit to a single column (60 mm

width, final size) or page (130 mm width) in the published paper. Authors should take care to ensure that symbols, labels, lines, etc. are large enough to allow reduction to a final size of *c.* 8 point, so that capital letters will be about 2 mm tall on publication. High contrast dots or line patterns are preferable to using different shades of grey, since they provide greater clarity on printing. Multiple graphs within one figure, should be marked with (A), (B), etc, and with an explanation for each provided in the figure heading. Authors should aim to ensure that there is no wasted space between multiple graphs.

Figures should **not be boxed** and tick marks must be on the **outside of the axes**. To make best use of space, you may need to rearrange parts of figures, for instance so that they appear side by side. Legends should furnish enough detail for figures to be understood without reference to the text. If symbols are straightforward (circles, squares, crosses or triangles), include them in the figure legend; otherwise they should be explained in the figure heading. Please keep the type size the same on legends and keys for all figures. Lettering should be in Garamond or Arial font with capitals used for the initial letter of the first word only. Bold lettering should not be used. Units of axes should appear in parentheses after the axis name.

If possible, please submit your artwork in electronic form. Large files (> 6MB) should be submitted on disk or CD, or via internet file transfer systems. Vector graphics (*e.g.* line artwork) should be saved in Encapsulated Postscript Format (EPS), and bitmap files (*e.g.* half-tones) in high resolution Tagged Image File format (TIF or TIFF). Graphs may also be submitted as Excel (.XLXS) files, and maps as Portable Document Format (.PDF) or Joint Photographic Experts Group (.JPEG) files. Ideally, figures should also be included within the text file at the end of the paper.

Scientific Names

Apply capitals as follows: Bewick's Swan, Pink-footed Goose *etc.*; but swans or geese. Follow an appropriate authority for common names, e.g. *Checklist of Birds in Britain and Ireland, Birds of the Western Palearctic* etc. Give the scientific (Latin) name of each species in full, in italics, at first mention in the main text, not separated by a comma or brackets, *e.g.* ...Red-throated Diver *Gavia stellata* was... . If there are many species, cite a check-list which may be consulted for authorities instead of listing them in the text. Do not give authorities for species cited from published references. Where appropriate, follow the Voous order of species.

Presentation of statistical information

Most statistical tests result in an estimate of the likelihood that a particular result could have arisen by chance. This probability is denoted by *P.* Authors are encouraged to follow the normal convention of indicating the probability of the result having arisen by chance and should be indicated by the use of < (less than) followed by the appropriate level (0.05, 0.02, 0.01, 0.001) taken from a set of statistical tables. Any result with a probability greater than 0.05 should be regarded as not significant and denoted by n.s. in the text, (not by $P > 0.05$). Authors are strongly recommended to follow the practice of using a null hypothesis before carrying out a test. In all cases, present the degrees of freedom, using a post-fix to the statistic symbol, for instance: χ^2_4, r_8, t_{28}.

Authors should ensure that the test they carry out is appropriate and the data are acceptable for the particular test. Ensure that the statistic is calculated correctly when small samples are involved (this includes the use of Yates' correction for the calculation of χ^2_1).

Examples of the style in which to present results are:

"... and the difference is significant ($\chi^2_1 = 6.9$, $P < 0.01$)."

"... the correlation between A and B is significant ($r_{28} = 0.79$, $P < 0.001$)."

"The difference between the samples is not significant ($t_{17} = 1.2$, n.s.)."

"Examination of the data using an ANOVA gives $F_{12,23} = 29.1$, $P < 0.001$."

Revision

All submitted manuscripts are subject to peer review, normally by at least two referees. If a manuscript is returned for revision, the corresponding author will be instructed to complete this exercise by a specific date; if an extension is required please contact the Editorial Assistant. It is expected that a manuscript accepted for publication in *Wildfowl* is in a form which is satisfactory to the author/s and will therefore not require changes at proof stage, other than the correction of errors occurring during the preparation for printing.

Pre-publication

Proofs will be sent to the correspondence author by e-mail, as an Acrobat PDF (portable document format) file. Acrobat Reader will be required in order to read this file. This software can be downloaded (free of charge) from the following web site: www.adobe.com/products/acrobat/readstep2.html

Only minor alterations may be accepted at this stage, unless approved by the Editor. Proofs must be returned to the editorial office by first class/air mail and by the date given. Alternatively, authors may use email to submit amendments providing the changes are clear and concise. The Editor reserves the right to correct the proofs, using the accepted version of the typescript, if the author's amendments are overdue and the journal would otherwise be delayed. Proofs should be checked very carefully. It is the corresponding author's responsibility to ensure that proofs are correct in every respect.

Reprints

Following publication, an electronic (PDF) copy of the paper will be sent to the corresponding author by e-mail. The PDF file will be sent to other authors of the paper on request. Papers will also be available for download from the WWT website at www.wwt.org.uk/wildfowl-journal. The complete back catalogue of *Wildfowl* is available on the Open Journal system at http://wildfowl.wwt.org.uk.

If you have any queries regarding the submission of papers to *Wildfowl*, please write to the Editor or send an e-mail to: wildfowl@wwt.org.uk

The Wildfowl & Wetlands Trust
Slimbridge, Gloucestershire GL2 7BT
United Kingdom
Tel: +44 (0)1453 891900
Fax: +44 (0)1453 891901 or 890827
E-mail: wildfowl@wwt.org.uk

WWT wetland centres

WWT centres offer a unique visitor experience where everyone can enjoy wetland habitats and their wildlife all year round in accessible and comfortable surroundings.

WWT Slimbridge
(Includes WWT Head Office)

Slimbridge
Gloucestershire GL2 7BT
United Kingdom
T +44 (0)1453 891900
F +44 (0)1453 890827
E wildfowl@wwt.org.uk
   membership@wwt.org.uk
http://www.wwt.org.uk

WWT Arundel

Mill Road, Arundel,
West Sussex BN18 9PB
T +44 (0)1903 883355

WWT Caerlaverock

Eastpark Farm, Caerlaverock,
Dumfriesshire DG1 4RS
T +44 (0)1387 770200

WWT Castle Espie

Ballydrain Road, Comber,
County Down BT23 6EA
T +44 (0)28 9187 4146

WWT London Wetland Centre

Queen Elizabeth's Walk, Barnes,
London SW13 9WT
T +44 (0)20 8409 4400

WWT Martin Mere

Burscough, Ormskirk,
Lancashire L40 0TA
T +44 (0)1704 895181

WWT National
Wetland Centre Wales

Llwynhendy, Llanelli,
Carmarthenshire SA14 9SH
T +44 (0)1554 741087

WWT Washington

Pattinson, Washington,
Tyne & Wear NE38 8LE
T +44 (0)191 416 5454

WWT Welney

Hundred Foot Bank, Welney,
Nr Wisbech, Norfolk PE14 9TN
T +44 (0)1353 860711

Photograph: Cannon-netting Greenland White-fronted Geese in Iceland during autumn 2017, by Stephanie Cunningham.

Photograph: Greenland White-fronted Goose in flight near Wexford, Ireland during spring 2013, by Chris Wilson.